Recipes
from a
Butcher's Wife

MARILYN STOYSICH

STOYSICH
PUBLISHING
OMAHA, NEBRASKA

Text and photographs © 2013 Marilyn Stoysich

All rights reserved. No part of this publication may be reproduced, stored in a retrieval system or transmitted in any form or by any means, electronic, mechanical, photocopying, recording, or otherwise, without the written permission of the publisher.

Paperback: ISBN 978-0-9886948-1-1
LCCN: 2012922621
Library of Congress Cataloging in Publication data on file with the publisher.

Send inquiries to the Publisher:
Stoysich Publishing
A Division of Frank Stoysich Meats

Design and publishing coordination: Concierge Marketing Inc.

Second Edition

Printed in the United States of America
10 9 8 7 6 5 4 3

Frank and Marilyn Stoysich

As a butcher's wife and a mother of four, I was probably best at the meat and potato entrées, usually adding a dash of some extra ingredient to come up with meals that were pleasing to my family.

I hope you enjoy this collection of my recipes, and I wish you the best for many enjoyable meals shared with your family.

INTRODUCTION

In 1960, Frank and I became proprietors of a small meat and grocery store formerly operated by his parents, Thomas and Elizabeth Stoysich. The business featured a variety of groceries as well as a good selection of meat products, and was reminiscent of several small mom and pop stores in the South Omaha neighborhood where I grew up. At that time, I never imagined that I would one day be the co-owner of a small retail business. But it seems that I was destined to become a "butcher's wife," and I soon came to enjoy this new found career, working side by side with my husband every day.

Due to my Polish and Czech heritage, I knew how to cook reasonably well before I was married, but now, because of my husband's Hungarian background, I was introduced to various new recipes. I prepared meals for two at first, but as the years went by and our children eventually numbered four, I began to try more ways to provide my family with wholesome meals. And, as our business evolved from fewer groceries to more meat and sausage items, I came up with new and diverse ways to include those meat products in my everyday menus.

When customers asked how to prepare a certain meal, I supplied suggestions and cooking methods for the meat items we sold. Printed recipes were offered on a regular basis. My ideas were well received and offered new cooking options for food preparation. Recipes featured basic dinners, Holiday entrées, outdoor grilling methods, and even some gourmet creations which earned special awards in Nebraska state meat competition.

Over the years, there have been numerous requests for a publication of my recipes. This cookbook will perhaps fill those requests and add to the enjoyment of family meals.

Marilyn Stoysich,
The "Butcher's Wife"

This Book is dedicated to Frank P. Stoysich, Sr.
1939 - 2011

And to my children
Frank Jr., Cathy, Tracey and Christine
For their continued support and encouragement

Red & White Store Prior to 1949

Tom's Market – 1951
Frank Stoysich, Sr. and
Thomas Stoysich

Frank Stoysich Meats – 1979
Left To Right: Frank Stoysich, Jr., Thomas
Stoysich and Frank Stoysich, Sr.

Frank Stoysich Meats – 1980, Store Interior – Meat Counter
Butchers Pictured: Leonard Gunia, Tim Conway

Marilyn And Frank Stoysich, Sr.
With National Awards – 1981

Marilyn Stoysich – 1983
Meat Product Demonstration

Original Building – Three Alarm Fire
December – 1985

Original Building Razed
June 29, 1986

Marilyn and Frank, Sr.
Grand Opening - 1987

Grand Opening – New Retail Store
February - 1987

Frank Stoysich Meats
Current Store Interior – 2012

Frank Stoysich Meats
Store Exterior - 2012

CONTENTS

INTRODUCTION III
DEDICATION V

Breakfast and Brunch 1
Appetizers 11
Soups 27
Salads 37
Meats—The Main Entrée 51
The Outdoor Grill 111
Fish and Seafood 125
Specialty Meat Entrées 133
Potatoes 141
Rice 155
Vegetables 161
Desserts 177
Sauce and Gravy 189

ALSO

Food Safety Guidelines 201
Cooking Instructions For Fresh Meats 203
Cooking Instructions For Smoked Meats and Sausage 205
Recipe Index 207

Breakfast and Brunch

Brunch is a meal served in between—anytime from approximately 10:00 a.m. to 2:00 p.m.—and can include standard breakfast items such as eggs, sausage, bacon or ham, as well as several other menu choices.

Brunch has always been a very popular Sunday offering at restaurants and in hotel dining rooms. The family Brunch is a wonderful way to begin a Holiday celebration at home.

Included in this section are some of my family's favorite recipes. The Apple Sausage Pancake is delicious, as is the Breakfast Quiche, which we have enjoyed at numerous Holiday gatherings.

APPLE SAUSAGE PANCAKE

Yield: 10 to 12 servings

INGREDIENTS:

1-1/2 pounds bulk pork sausage

3 to 4 Jonathan or Granny Smith apples

1-1/2 cups water

3 Tbsp. granulated sugar

1/4 tsp. ground cinnamon

2 cups buttermilk pancake mix

1 cup milk

5 eggs, beaten

4 Tbsp. butter or margarine, softened

1/2 cup granulated sugar

2 tsp. cinnamon

BROWN AND CRUMBLE THE PORK SAUSAGE. Eliminate excess fat and cool.

PEEL, CORE AND SLICE THE APPLES.

COOK APPLE SLICES with the water, 3 Tbsp. sugar and 1/4 tsp. cinnamon for approximately 20 minutes on medium heat. Drain water and cool.

IN A BOWL, combine the pancake mix, milk and eggs. Mix well.

SPRAY TWO 10-INCH GLASS PIE PLATES with vegetable coating spray. In each pie plate, spread 2 Tbsp. butter or margarine and arrange half of the apple slices, and half of the cooked pork sausage.

MIX TOGETHER the remaining sugar and cinnamon and sprinkle 2 Tbsp. of this mixture over the apples and sausage in each pie plate.

DIVIDE PANCAKE BATTER IN HALF and pour over all in each pie plate. Top each with 2 Tbsp. of the sugar and cinnamon mixture.

BAKE IN A 325° OVEN for 12 to 15 minutes or until golden brown. Cut apple sausage pancakes into wedges and serve with whipped butter and a variety of syrups and fruit preserves.

Bacon Brunch Muffins

Yield: 10 to 12 servings

INGREDIENTS:

6 slices hickory smoked bacon

1 cup sifted all-purpose flour

1-1/4 cups yellow cornmeal

2-1/2 tsp. double acting baking powder

2 Tbsp. granulated sugar

3/4 tsp. salt

1 cup milk

1 egg, beaten

3 Tbsp. reserved bacon drippings

IN A SKILLET, cook bacon and set on paper towels to cool. Coarsely chop the bacon.

Reserve the bacon drippings.

COMBINE IN A BOWL the flour, cornmeal, baking powder, sugar and salt. Add the milk, egg and bacon drippings and mix well. Stir in the chopped bacon.

LINE A MUFFIN PAN with paper liners and fill each about two thirds full.

BAKE IN A 325° to 350° oven approximately 25 minutes.

BREAKFAST QUICHE

Yield: 6 servings

INGREDIENTS:

8 or 10 slices hickory smoked bacon, fried crisp, drained and coarsely chopped

1 cup shredded Swiss cheese

1/3 cup finely chopped onion

2 cups milk or half & half

1/2 cup Bisquick baking mix

4 or 5 beaten eggs

1/4 tsp. salt

1/8 tsp. black pepper

1/8 tsp. ground nutmeg

LIGHTLY GREASE A 10-INCH PIE PLATE. Sprinkle the bacon, cheese and onion evenly in the bottom of the pie plate. Mix in a blender the milk or half & half, Bisquick, eggs and seasonings. Blend on high speed about one minute.

POUR MIXTURE INTO THE PIE PLATE OVER BACON, CHEESE AND ONION. Bake at 350° for about 45 to 50 minutes, or until golden brown and knife inserted into the quiche comes out clean. Let stand a few minutes before serving.

BRUNCH CASSEROLE

Yield: 6 servings

INGREDIENTS:

6 slices French toast, fresh or frozen OR 6 frozen waffles

12 medium slices Canadian bacon

6 to 8 hard-boiled eggs, sliced

salt and pepper to taste

1 10 oz. can cheddar cheese soup

1/3 cup milk

1 2 oz. jar sliced pimento, drained

1/2 cup chopped green pepper

1/2 cup shredded mild cheddar cheese

COAT BOTTOM OF A 9x13 GLASS BAKING DISH with vegetable coating spray. If using frozen French toast or waffles, toast these slightly. Place the toast or waffles in the baking dish and top with Canadian bacon. Next, add the sliced, hard cooked eggs and salt and pepper to taste.

IN A SMALL BOWL mix canned soup with milk, pimento and green pepper. Spoon this mixture over all ingredients in the baking dish, covering them evenly.

TOP WITH SHREDDED CHEESE and bake in a 325° to 350° oven for approximately 30 minutes.

Eggs-Benedict Canadian

Yield: 6 servings

INGREDIENTS:

6 medium slices Canadian bacon

3 English muffins, split

butter

6 poached eggs*

salt and pepper to taste

1 pkg. hollandaise sauce mix, prepared according to package directions

paprika to taste

IN A SKILLET heat the slices of Canadian bacon.

TOAST ENGLISH MUFFIN HALVES and butter lightly.

Top each muffin half with 1 slice of Canadian bacon.

PLACE ONE POACHED EGG ON CANADIAN BACON and add salt and pepper.

Top with prepared hollandaise sauce. Add a dash of paprika.

*To poach eggs, bring water to a boil in a deep pan or skillet on top of the stove. Reduce to a slow simmer and crack each room temperature egg into a small bowl or dish. Carefully add eggs (one at a time) to the water and cook for approximately 2 to 3 minutes. The whites will cook faster than the yolks.

Ham and Cheddar Cups

Yield: About 20 Ham and Cheddar Cups

INGREDIENTS:

1/2 pound hickory smoked bacon, cooked and chopped—set aside on paper towels

1/2 cup diced hickory smoked boneless ham

2 cups flour

1/4 cup granulated sugar

2 tsp. baking powder

1 tsp. salt

1/4 tsp. coarse ground black pepper

6 eggs, beaten

1 cup milk

1/2 cup shredded cheddar cheese

1 small onion, chopped fine

IN A MIXING BOWL COMBINE flour, sugar, baking powder, salt and pepper. Combine eggs with milk and stir into dry ingredients. Add bacon, ham, cheese and onion.

FILL WELL-GREASED MUFFIN CUPS about 3/4 full. Bake in a 325° to 350° oven for approximately 45 minutes or until toothpick inserted in muffin comes out clean. Cool about 10 minutes before moving to a wire rack.

Note:
Paper liners in muffin pan are not recommended.

Hash Brown and Egg Bake

Yield: about 8 servings

INGREDIENTS:

8 or 10 slices hickory smoked bacon fried crisp and chopped—set aside on paper towels

1 32 oz. pkg. frozen hash browns, thawed

1-1/2 cups shredded cheddar cheese, divided

1/4 to 1/2 tsp. salt

black pepper to taste

8 eggs

2 cups milk

1/4 tsp. paprika

IN A BOWL, COMBINE HASH BROWNS, BACON, half the cheese and the salt and pepper. Spoon into a greased 9x13-inch glass baking pan.

IN ANOTHER BOWL, BEAT EGGS AND ADD MILK—mix until smooth. Pour this over the hash brown mixture.

SHAKE ON PAPRIKA AND BAKE, UNCOVERED, in a 325° oven for about 30 minutes. Top with the remaining cheese and continue baking for another 10 or 15 minutes.

Note:
This breakfast casserole may be prepared ahead and refrigerated overnight. Remove from refrigerator about 30 minutes before baking.

Sausage Cake

Yield: 3 Sausage Cakes

The following is actually a recipe for an old-fashioned German fruit cake. The bulk sausage, which is not cooked before mixing, serves as the shortening for this cake.

INGREDIENTS:

- 3 eggs, well beaten
- 1 pound bulk pork sausage
- 2 cups granulated sugar
- 3 cups flour
- 2 tsp. baking soda
- 1 cup hot coffee or water
- 1 8 oz. box chopped dates
- 1 cup dark or golden raisins
- 1 cup chopped walnuts or pecans
- 2 tsp. ground cinnamon
- 1 tsp. allspice
- 1 tsp. ground nutmeg

COMBINE EGGS, SAUSAGE, SUGAR AND FLOUR in a mixing bowl. Dissolve baking soda in hot coffee or water and add to sausage mixture. Add remaining ingredients and mix thoroughly.

PLACE BATTER IN GREASED LOAF PANS and bake in a 325° to 350° oven for about 35 to 45 minutes, or until toothpick inserted into sausage cake comes out clean.

Serve warm.

Store leftover cake in the refrigerator.

Appetizers

    Appetizers, whether plain or fancy, are a type of "first course" before the first course of a meal. They serve as a way to stimulate the appetite, or to satisfy guests during the cocktail hour and before dinner is served. The Italians refer to appetizers as 'Antipasto'—the French word for appetizers is 'Hors d'oeuvres.'

    My memory of appetizers at family gatherings years ago was one of very simple offerings such as pickles and olives, chips and dip, crackers and cheese, and on a rare and special occasion—shrimp cocktail.

    An appetizer buffet can feature seafood, creamy spreads, meatballs and other finger foods—and can even be a meal in itself. In fact, for an occasional Holiday gathering, the dinner menu for my family has consisted only of various Hors d'oeuvres and appetizers, instead of a several course meal.

APPETIZER RIBS

Yield: about 20 appetizers

INGREDIENTS:

1-1/2 slabs pork loin back ribs, sawed in half lengthwise

BARBEQUE SAUCE:

1 15 oz. can tomato sauce

1/4 cup cider vinegar

1/3 cup catsup

3 Tbsp. Worcestershire sauce

1/2 cup brown sugar

2 Tbsp. paprika

1 tsp. garlic salt

1/2 tsp. black pepper

1 tsp. chili powder

Tabasco sauce as desired

Cut rib halves into two-rib sections, place in a covered roaster pan and bake in a 325° oven approximately 45 minutes. Excess juice or fat should be eliminated, with a small amount of liquid remaining in roaster.

MIX TOGETHER BARBEQUE SAUCE IN THE ORDER GIVEN, adding as much or as little Tabasco sauce as preferred. Begin basting ribs in roaster pan with the barbeque sauce and return to oven (uncovered.)

LOWER OVEN TEMPERATURE TO 300° and bake an additional 30 to 45 minutes or until rib sections are tender. Baste frequently with sauce during this portion of baking time.

Artichoke Spinach Spread

Yield: 6 cups

INGREDIENTS:

1 14 oz. can artichoke hearts, drained and chopped

1 10 oz. pkg. frozen chopped spinach, thawed and drained

1/2 tsp. minced garlic

2 cups mayonnaise

2 cups shredded Parmesan cheese

Mix all ingredients until well blended. Pour into a casserole dish and bake uncovered at 325° for about 45 minutes.

Serve with crackers

Stuffed Cherry Tomatoes

Yield: 24 tomatoes

INGREDIENTS:

24 fresh cherry tomatoes

CRAB AND SHRIMP FILLING

2 8 oz. pkgs. Philadelphia cream cheese, softened

1 6 oz. can crabmeat, drained

1 4 oz. can tiny shrimp, drained and rinsed

2 Tbsp. finely minced green onion

1 tsp. dill weed

Lawry's seasoned salt

WASH CHERRY TOMATOES, and with a sharp paring knife remove stem, and cut a small slice from the top of each so that pulp and seeds may be removed. Place tomatoes upside down on paper towels.

MEANWHILE, MIX TOGETHER CRAB AND SHRIMP FILLING.

ARRANGE CHERRY TOMATOES on a bed of leaf lettuce on a serving platter. Spoon the crab-shrimp filling into a pastry bag with large piping tube. Fill each tomato with seafood mixture and shake on additional dill weed.

Chili-Nacho Dip

Yield: about 6 cups

INGREDIENTS:

1 pound ground beef

1/4 cup diced onion

2 16 oz. pkgs. Velveeta cheese

1 15 oz. can chili (no beans)

1 4 oz. can chopped black olives

1/2 cup salsa or picante sauce

1/2 tsp. ground cumin

BROWN AND CRUMBLE GROUND BEEF AND ELIMINATE EXCESS FAT.

Sauté the onion. Cut up and slowly melt the Velveeta cheese in a saucepan on top of the stove. Add ground beef, onion, canned chili, chopped olives, salsa or picante sauce and cumin. Heat slowly while stirring.

Serve with tortilla chips

CHIPPED BEEF DIP

Yield: about 4 cups

INGREDIENTS:

2 8 oz. pkgs. Philadelphia cream cheese

1/2 cup sour cream

1/2 pound dried beef, finely chopped

1 small green pepper, diced

1 bunch green onions, sliced

2 Tbsp. Worcestershire sauce

1/2 tsp. coarse ground black pepper

Lawry's seasoned salt to taste

SOFTEN CREAM CHEESE AND COMBINE INGREDIENTS IN ORDER GIVEN. Refrigerate for at least an hour and serve with crackers. If mixture is too thick add a tablespoon or two of milk or half & half.

Cucumber Sandwiches

Yield: 21 sandwiches

INGREDIENTS:

1 loaf cocktail rye bread

Philadelphia cream cheese

sliced fresh cucumbers

dill weed

SPREAD CREAM CHEESE on individual pieces of cocktail bread. On each, place a cucumber slice and top with a sprinkling of dill weed.

ARRANGE SANDWICHES on a platter. Use sheets of waxed paper or parchment doilies to separate layers.

DEVILED EGGS

Yield: 24 halves

INGREDIENTS:

1 dozen hard-boiled eggs, shelled and cooled

4 additional hard-boiled eggs*

1/2 cup mayonnaise

2 Tbsp. finely minced onion

1 Tbsp. white vinegar

1/4 tsp. dry mustard

2 Tbsp. sweet pickle relish

salt and pepper to taste

paprika to taste

CUT EGGS IN HALF LENGTHWISE. Carefully remove yolks, place in a bowl and mash them. Blend in the mayonnaise and all remaining ingredients. Mixture should be stiff enough to fill yolk halves with a pastry tube. (Use a large decorating tip.)

DISPLAY DEVILED EGGS ON A PLATTER LINED WITH LEAF LETTUCE. Top with a sprinkling of paprika and a sprig of fresh parsley if desired.

Note:
*It is always a good idea to cook a few additional eggs to allow for breakage and imperfections.

Liver Spread

Yield: 2 cups

INGREDIENTS:

1 pound chicken livers
butter or margarine
1/2 cup sliced onion
1 tsp. garlic powder
1/2 tsp. black pepper
1 tsp. salt or seasoned salt
a pinch of paprika

SAUTÉ LIVERS AND ONION 8 to 10 minutes in a skillet with a small amount of butter or margarine, until livers are browned on the outside, but slightly pink on the inside.

IN A BOWL, COMBINE LIVERS AND ONION WITH seasonings.

PUT MIXTURE IN BLENDER OR FOOD PROCESSOR, adding a small amount of water, and blend until smooth. Refrigerate before serving.

Serve with crackers.

Swedish Meatballs

Yield: 30 or more appetizer meatballs

INGREDIENTS:

2 pounds ground beef

1 egg, beaten

1/2 pound bulk pork sausage

1/2 tsp. salt

1/2 cup finely minced onion

1/2 tsp. black pepper

3/4 cup bread crumbs

COMBINE INGREDIENTS, ADDING A LITTLE WATER TO MOISTEN. Form into one-inch balls and bake on a greased pan or cookie sheet in a 325° oven for approximately 20 to 30 minutes. Turn to brown.

EASY SWEDISH SAUCE:

2 10 oz. cans cream of mushroom soup

small amount of milk or sour cream

1/2 tsp. ground nutmeg

1/4 tsp. ground allspice

COMBINE AND HEAT SAUCE INGREDIENTS and add the meatballs. Keep warm in crock pot or chafing dish.

Stuffed Mushrooms

Yield: 32 appetizers

INGREDIENTS:

32 large mushrooms

3/4 pound bulk pork sausage

1/2 Tbsp. garlic powder

1/2 to 3/4 cup bread crumbs

2/3 cup grated Parmesan cheese

fresh snipped parsley or parsley flakes

salt and pepper to taste

2/3 cup melted butter or margarine

WASH MUSHROOMS AND REMOVE STEMS. Chop stems. Brown and crumble pork sausage. Add chopped mushroom stems, garlic powder, bread crumbs, Parmesan cheese, fresh snipped parsley or parsley flakes and salt and pepper to taste.

PLACE MUSHROOM CAPS ON BAKING SHEET AND FILL EACH MUSHROOM CAP with about one tablespoon of filling and brush with melted butter or margarine. Broil until bubbly - about 3 to 5 minutes.

PICKLE ROLL-UPS

Yield: about 48 appetizers

INGREDIENTS:

8 medium dill pickles

8 slices of corned beef - medium thickness

8 oz. Philadelphia cream cheese, softened

DRY DILL PICKLES WITH PAPER TOWELS to eliminate juice. Spread softened cream cheese on corned beef, and put one pickle on each slice. Roll tightly.

WHEN ALL ARE ASSEMBLED, freeze them for about 10 or 15 minutes to firm. Then slice each into about 8 pieces. (End pieces make good tasting samples.)

DISPLAY THE PICKLE ROLL-UPS on a serving tray, separating layers with sheets of waxed paper or parchment doilies.

Sausage-Cheese Puffs

Yield: 2 dozen puffs

INGREDIENTS:

1 pound bulk mild Italian sausage

3 cups Bisquick baking mix

4 cups shredded cheddar cheese

3/4 cup water

BROWN AND CRUMBLE SAUSAGE AND DRAIN FAT - set aside to cool. In a bowl, combine Bisquick and cheese. Stir in water and mix with a fork until moistened. Mix in sausage, and shape into one and one-half inch balls.

BAKE TWO INCHES APART ON A BAKING SHEET (which has been lightly sprayed with vegetable coating) in a 325° to 350° oven for 12 to 15 minutes or until puffy and golden brown.

Note:
Sausage-Cheese Puffs may be prepared ahead and frozen. They need not be thawed before reheating (about 7 to 10 minutes at 300° or until heated through.)

SAUSAGE AND POTATO APPETIZERS

Yield: about 36 appetizers

INGREDIENTS:

6 mini-Polish sausages or natural casing wieners

2 cups prepared mashed potatoes, cooled

1 beaten egg

3 Tbsp. minced onion

1 Tbsp. parsley flakes (or fresh minced parsley)

1/2 tsp. garlic salt

1/4 tsp. black pepper

2 cups seasoned bread crumbs

1 beaten egg mixed with 1/2 cup milk

vegetable oil

CUT EACH SAUSAGE OR WIENER into 6 bite-sized pieces.

MIX MASHED POTATOES WITH egg, onion, parsley flakes or fresh parsley, garlic salt and black pepper. Press potato mixture (one tablespoon or less) around each sausage piece. This becomes easy after doing a few—using too much potato mixture makes these larger than appetizer size. A small amount of flour may be added to potato mixture to firm the texture.

ROLL SAUSAGE BALLS IN THE SEASONED BREAD CRUMBS, then in egg-milk mixture, and roll again with bread crumbs.

DEEP FRY IN HOT OIL (about 360°) until golden brown. Drain on paper towels before serving.

Shrimp Spread

Yield: 4 cups

INGREDIENTS:

2 4 oz. cans tiny shrimp, drained and rinsed

2 8 oz. containers whipped cream cheese

3/4 cup finely chopped celery

1/2 cup finely chopped onion

2 tsp. Paul Prudhomme's Seafood Magic seasoning

1/2 cup sour cream

1/2 tsp. garlic powder

1 Tbsp. dried parsley flakes

cocktail sauce for topping

MIX INGREDIENTS IN ORDER GIVEN. Shrimp spread should be firm enough to mound onto serving platter. Top with bottled cocktail sauce or homemade cocktail sauce, on page 192 of "Sauce and Gravy" Section. Chill before serving. Arrange crackers around edge of platter.

Sweet and Sour Mini-Bites

Yield: 40 or more appetizers

INGREDIENTS:

10 mini-Polish sausages

SWEET-SOUR SAUCE:

1 cup pineapple juice

1 cup chicken broth

1/2 cup brown sugar

1/3 cup cider vinegar

1/4 cup catsup

2 Tbsp. soy sauce

2 Tbsp. cornstarch

1/2 cup sliced green onion

IN A SAUCEPAN HEAT THE PINEAPPLE JUICE, CHICKEN BROTH, SUGAR AND VINEGAR. Mix cornstarch with catsup and soy sauce and add to mixture in saucepan. Heat and stir until thick. Add green onion and lower heat.

CUT EACH MINI-POLISH SAUSAGE INTO 4 OR 5 BITE-SIZED PIECES. Combine with sweet-sour sauce and keep warm in a crockpot or chafing dish.

SOUPS

One of my fondest memories is that of the chicken soup made by my Polish grandmother—she prepared this dish only one way. She simmered the cut-up chicken with vegetables in a stock pot until the broth was rich and the chicken was tender. The chicken pieces were transferred to a skillet and "browned" for a type of sautéed chicken—then noodles were added to the soup mixture. Thus—two meals—delicious chicken soup and 'fried chicken.' My recipe for chicken soup, although not like my grandmother's, is included together with several others.

Oxtail soup was one of my children's and grandchildren's favorites—they referred to it as "ox-bone" soup. Great Northern Bean Soup can't be beat on a cold winter's day, and the Quick Clam Chowder is a soup I came up with using canned clams and a packaged vegetable soup mix.

BROCCOLI-CAULIFLOWER CHEESE SOUP

Yield: 8 to 10 servings

INGREDIENTS:

5 cups small broccoli florets

5 cups small cauliflower florets

1 cup finely chopped onion

6 Tbsp. butter

1/2 cup flour

6 cups chicken broth

6 cups whole milk

2-1/2 tsp. Worcestershire sauce

4 cups shredded American cheese

4 cups shredded sharp cheddar cheese

1/2 cup shredded jalapeño pepper cheese

STEAM BROCCOLI AND CAULIFLOWER FLORETS UNTIL JUST TENDER. Set aside.

IN A 6 TO 8 QUART STOCKPOT, cook onion in butter. Blend in the flour and combine well with the butter and onions. Stir in the chicken broth, milk and Worcestershire sauce and cook uncovered on low to medium heat, stirring often, until mixture begins to thicken.

REDUCE HEAT TO LOW AND ADD ALL OF THE SHREDDED CHEESE, stirring until cheese is melted.

ADD STEAMED BROCCOLI AND CAULIFLOWER to the soup and continue cooking for about 20 more minutes.

Chicken Soup

Yield: 6 to 8 servings

INGREDIENTS:

2 whole bone-in chicken breasts, split

1 to 2 Tbsp. granulated chicken bouillon

1 pound carrots, peeled and sliced

1-1/2 to 2 cups chopped celery

3 to 4 potatoes, peeled and cut in small pieces

1 medium onion, chopped

1 15 oz. can diced tomatoes (optional)

salt and pepper to taste

1 16 oz. pkg. frozen egg noodles

parsley flakes to taste

PUT CHICKEN BREASTS IN HEAVY KETTLE WITH WATER TO COVER. Bring to a moderate boil, and cook for 1 to 1-1/2 hours, or until meat thermometer registers 165° to 170° when inserted into the chicken. Remove any impurities from the broth during this cooking time.

ADD CHICKEN BOUILLON TO THE BROTH, and more water if needed.

TAKE CHICKEN OUT OF THE KETTLE AND ALLOW TO COOL. Discard skin, remove meat from bone, cube or dice chicken, and return to the broth. At this time, add the vegetables, salt and pepper, and simmer for about 15 minutes before adding frozen noodles. Continue simmering for 15 to 20 minutes or until vegetables and noodles are done. Canned tomatoes give this soup a wonderful flavor, and parsley flakes are also a nice touch!

Makes a generous amount.

Creamy Potato Soup

Yield: 6 to 8 servings

INGREDIENTS:

2 14 oz. cans chicken broth

2 cups water

5 potatoes, peeled and cut in small chunks

6 Tbsp. butter or margarine

1/2 cup finely chopped celery

1/2 cup chopped onion

1/2 cup diced carrots

3/4 cup flour

2 cups whole milk

1 tsp. celery salt

1 tsp. coarse ground black pepper

1 cup sour cream

1/2 cup additional chicken broth

2 cups shredded white American cheese

BRING CHICKEN BROTH AND WATER TO A SIMMERING BOIL in kettle or stock pot on top of the stove. Add the potatoes and cook until nearly tender – about 15 minutes.

MEANWHILE, IN A LARGE SKILLET MELT butter or margarine and sauté the vegetables until nearly tender. Add the flour, milk, celery salt and black pepper. Cook on low heat, stirring until mixture thickens.

ADD VEGETABLE MIXTURE TO THE POTATOES AND BROTH. Continue simmering on low another 15 to 20 minutes.

IN A SMALL BOWL, COMBINE the sour cream with additional broth. Add to the soup.

Stir in the shredded cheese and simmer a few more minutes until cheese melts.

Top with minced fresh parsley or parsley flakes.

Favorite Vegetable Soup

Yield: about 10 servings

INGREDIENTS:

12 cups water

2 Tbsp. granulated chicken bouillon

1 14 oz. can diced tomatoes

1 medium onion, chopped fine

2 tsp. granulated sugar

1 tsp. coarse ground black pepper

1 tsp. ground oregano

2 tsp. ground cumin

1 12 oz. can tomato paste

6 carrots, peeled and sliced

1 small head cabbage, core removed and chopped

2 small zucchini squash, sliced

1-1/2 cups chopped celery

1 16 oz. pkg. frozen green beans

2 medium potatoes, peeled and diced

2 additional cups of water (approximate)

parsley flakes

IN STOCK POT OR SOUP KETTLE bring to a slow simmering boil the 12 cups of water and chicken bouillon.

ADD TO THE STOCKPOT: tomatoes, onion, sugar, black pepper, oregano and cumin.

Simmer on low for about 45 minutes and then add the tomato paste, carrots, cabbage, zucchini, celery, green beans, potatoes and additional water.

CONTINUE SIMMERING ON LOW UNTIL ALL VEGETABLES ARE TENDER. Shake on parsley flakes and more seasoning as preferred.

Makes a generous amount.

Great Northern Bean Soup

Yield: 6 to 8 servings

INGREDIENTS:

1 smoked ham shank, cut into thirds

1 16 oz. pkg. dry great northern beans

1 medium onion, chopped

3 carrots, peeled and diced

5 or 6 ribs celery, diced

1 8 oz. can tomato sauce

coarse ground black pepper to taste

Soak beans in water at least six hours or overnight. Drain and rinse beans and place in stock pot with water to cover.

ADD HAM SHANK PIECES AND BRING TO A MODERATE BOIL. After about a half-hour, reduce heat and remove impurities from the top being careful not to remove beans. Add black pepper and onion and continue to simmer slowly for 2 to 3 hours. When beans are nearly tender, add diced carrots, celery, tomato sauce and more water if needed.

CONTINUE COOKING BEAN SOUP on low about an hour longer. At this time ham shank portions may be taken out of stock pot and meat removed from bone, diced, and returned to stock pot. A cup or more of broth may be removed from stock pot and thickened with a little cornstarch (first mixed in cold water), then returned to the soup.

Delicious with rye bread or corn bread and a green salad.

Makes a generous amount of soup.

CROCK POT INSTRUCTIONS:

Prepare as above, cover, and cook on low for 7 to 8 hours.

Hearty Beef Soup

Yield: 8 to 10 servings

INGREDIENTS:

1-1/2 to 2 pounds beef shank meat

1 14 oz. can stewed tomatoes

1 large onion, chopped

salt and pepper to taste

2 cups sliced celery

6 to 8 carrots, peeled and sliced

4 to 5 potatoes, peeled and cut in small pieces

1 pkg. frozen green beans (optional)

1 12 oz. pkg. frozen egg noodles

1 to 2 Tbsp. granulated beef bouillon

PLACE MEAT IN HEAVY STOCK POT OR KETTLE ON TOP OF THE STOVE. Add water to within 3 to 4 inches of top of kettle. Bring to a boil and simmer about 45 minutes. Remove impurities once or twice from broth, and continue cooking over low to moderate heat.

NEXT ADD tomatoes, onion, salt and pepper. Cover and continue simmering over low to moderate heat approximately two more hours or until meat is tender. Stir once or twice during this cooking time, adding water if necessary. Remove the beef shank and allow to cool slightly. Cut meat away from the bone, trimming excess fat if needed. Cube the meat and return to broth.

ADD the celery, carrots, potatoes, green beans and frozen noodles.

ADD BEEF BOUILLON and additional water if needed and simmer another 20 to 30 minutes. Shake on additional black pepper if desired and top with fresh snipped parsley or parsley flakes.

Makes a generous amount. Serve with hot crusty bread and a tossed salad.

OXTAIL SOUP

Yield: 8 to 10 servings

INGREDIENTS:

3 pounds oxtails

1 14 oz. can tomatoes, stewed or diced

1 medium onion, chopped

1 to 2 Tbsp. granulated beef bouillon

salt and pepper to taste

6 or 8 carrots, peeled and sliced

2 cups sliced celery

4 or 5 potatoes, peeled and cut in chunks

3 or 4 parsnips, peeled and sliced

1 cup quick cooking barley

1 12 oz. pkg. frozen egg noodles

PLACE OXTAILS IN HEAVY KETTLE ON TOP OF THE STOVE. Add water to cover, and begin simmering to a medium boil. Cook for about 1-1/2 hours, removing impurities as they come to the top.

ADD tomatoes, onion, beef bouillon, salt, pepper and additional water if needed.

CONTINUE SIMMERING SLOWLY— at least another hour or until oxtails are tender.

NEXT ADD carrots, celery, potatoes, parsnips, barley and frozen noodles.

ADD ADDITIONAL WATER IF NEEDED and simmer another 20 to 30 minutes. Before serving, top with parsley flakes.

Makes a generous amount.

Quick Clam Chowder

Yield: 6 to 8 servings

INGREDIENTS:

1 1.4 oz. pkg. Knorr vegetable soup mix

4 cups water

2 cans minced clams, in juice

1 cup fresh sliced mushrooms

coarse ground black pepper to taste

3 10 oz. cans cream of potato soup

IN A 3 QUART SAUCEPAN combine Knorr vegetable soup mix with water. Bring to a boil, then simmer on low (uncovered) for about 15 minutes.

ADD THE MINCED CLAMS with juice, the sliced mushrooms and black pepper. Continue simmering over low heat.

IN A MEDIUM SIZED BOWL, mix the cans of potato soup until smooth and add to the saucepan. Stir while simmering another 10 to 15 minutes.

Serve with oyster crackers or saltines.

SALADS

    A salad is most often served before a meal, but it can also be a complete meal. Salads can consist of a variety of leafy greens and vegetables, with seafood, meat and other ingredients. There are also fruit salads, dessert salads, and pasta salads made with dressings or mayonnaise.

    Our family always enjoyed salads to accompany a meal, and I'm sharing some of our favorites, such as Macaroni Shell Salad, Seven Layer Salad, and an Orange Dessert Salad.

    My Sea Foam Salad is very much like the Jell-O salad which was served at parish dinners and wedding receptions many years ago. In fact, it was at just such a dinner that I first met the 'Butcher.'

Cabbage Slaw

Yield: 4 to 6 servings

INGREDIENTS:

1 medium head green cabbage, shredded

1 green pepper, chopped

1 cup canned crushed pineapple, drained

1/2 cup sour cream

1/2 cup mayonnaise

1 tsp. granulated sugar

1/2 tsp. celery seed

COMBINE the cabbage, green pepper and drained pineapple.

MIX TOGETHER the sour cream, mayonnaise, sugar and celery seed.

ADD DRESSING TO CABBAGE SLAW MIXTURE. Refrigerate at least one hour before serving.

Cottage Cheese Salad

Yield: 1 serving

INGREDIENTS:

whole leaves of iceberg lettuce

1/2 cup or more cottage cheese

canned peach or pear halves

a dollop of mayonnaise

shredded cheddar cheese

PLACE LETTUCE LEAVES on a salad plate. Add cottage cheese, peach or pear halves and top with mayonnaise and a sprinkling of cheddar cheese.

Cran-Raspberry Jell-O Salad

Yield: 6 to 8 servings

INGREDIENTS:

2 3 oz. pkgs. raspberry Jell-O

1-1/2 cups boiling water

1 cup cold water

2 cups fresh cranberry sauce or 1 14 oz. can whole berry cranberry sauce

1-1/2 cups fresh ground raw cranberries

DISSOLVE JELL-O in 1-1/2 cups boiling water. Add 1 cup cold water. Stir and refrigerate.

WHEN JELL-O IS PARTIALLY SET, add cranberry sauce and fresh ground cranberries. Pour into a ring mold which has been sprayed with a vegetable coating spray. Chill until firm. Transfer carefully from mold to a platter garnished with endive or leaf lettuce.

Cucumber Salad

Yield: 4 to 6 servings

INGREDIENTS:

3 cucumbers

1 tsp. salt

ice water

1 medium onion, sliced

1 pkg. Good Seasons Italian salad dressing mix, prepared

1/4 cup more white vinegar

coarse ground black pepper to taste

WASH AND PEEL CUCUMBERS, leaving strips of green for color. Slice and put them in salted ice water and refrigerate at least an hour. Drain water and combine cucumbers with onion slices in a bowl.

MIX GOOD SEASONS ACCORDING TO PACKAGE DIRECTIONS adding an additional 1/4 cup white vinegar. Pour dressing over the cucumbers and onions and add black pepper.

IF DESIRED, top with additional dry Good Seasons mix.

Chill before serving.

Cucumbers and Sour Cream

Yield: 4 to 6 servings

INGREDIENTS:

3 cucumbers

1 medium onion, sliced

1 tsp. salt

ice water

1 cup sour cream

1 Tbsp. granulated sugar

2 Tbsp. white vinegar

1/4 tsp. paprika

1/4 tsp. dill weed

coarse ground black pepper to taste

WASH AND PEEL CUCUMBERS, leaving strips of green for color. Slice and put them in salted ice water and refrigerate at least an hour. Drain water and combine cucumbers with onion slices in a bowl.

MIX TOGETHER the sour cream, sugar and vinegar.

Pour dressing over cucumbers and onions and top with paprika, dill weed and black pepper.

VARIATION:

Substitute sliced onion with diced red onion, and add sliced cherry or grape tomatoes.

Macaroni Shell Salad

Yield: about 8 servings

INGREDIENTS:

1 24 oz. pkg. shell macaroni, cooked, drained and cooled

6 or 8 hard-boiled eggs, chopped

1 medium onion, finely chopped

1 1/2 cups chopped celery

2 cucumbers, peeled, seeded and chopped

2 firm ripe tomatoes, cut in small pieces

1 4 oz. can chopped ripe olives

1/2 cup sweet pickle relish, juice drained

2 cups Miracle Whip salad dressing

2 Tbsp. French's regular mustard

salt and pepper to taste

dill weed or dried parsley flakes, as preferred

IN A LARGE BOWL COMBINE the macaroni, eggs, onion, celery, cucumbers, tomatoes, olives, and pickle relish.

MIX TOGETHER the Miracle Whip salad dressing, mustard and seasonings. Pour this mixture over the macaroni and all other ingredients and stir to combine.

Chill before serving.

ORANGE DESSERT SALAD

Yield: 6 to 8 servings

INGREDIENTS:

2 3 oz. pkgs. orange Jell-O

2 cups boiling water

6 oz. frozen orange juice concentrate, thawed

2 11 oz. cans mandarin orange sections, drained

1 15 oz. can crushed pineapple, drained

1 3 oz. pkg. instant vanilla or lemon pudding

3/4 cup milk

8 oz. Cool Whip or whipped topping

1 cup finely chopped English walnuts

IN A BOWL dissolve Jell-O in boiling water. Add orange juice concentrate, mix well and place in refrigerator for about 30 minutes.

WHEN MIXTURE BEGINS TO SET UP, ADD mandarin oranges and pineapple.

TRANSFER JELL-O MIXTURE to a glass serving dish and return to refrigerator.

MIX TOGETHER the instant pudding and milk and stir until partially set. Combine pudding mixture with Cool Whip and spread on chilled Jell-O. Sprinkle the walnuts on top. Chill at least 2 hours before serving.

Quartered Lettuce

Yield: 4 to 6 servings

INGREDIENTS:

4 to 6 wedges iceberg lettuce

1000 ISLAND DRESSING:

1 cup real mayonnaise

1/2 cup bottled chili sauce

1 Tbsp. finely minced onion

1 Tbsp. chopped green pepper

2 hard-boiled eggs, chopped

Place lettuce wedges on individual salad plates, and top with 1000 Island dressing.

FOR THE DRESSING: IN A BOWL MIX mayonnaise with chili sauce, onion, green pepper and hard-boiled eggs.

Garnish plates with a few cherry tomatoes.

SEA-FOAM SALAD

Yield: 8 or more servings

INGREDIENTS:

2 3 oz. pkgs. regular lime Jell-O

2 cups boiling water

1-1/3 cups cold water

1 24 oz. carton small curd cottage cheese

2 15 oz. cans crushed pineapple, drained

IN A DISH, DISSOLVE LIME JELL-O IN BOILING WATER. ADD COLD WATER.

Place dish of Jell-O in refrigerator until partially set, about 30 to 45 minutes, stirring once or twice during this time. Then mix in the cottage cheese and pineapple.

RETURN TO REFRIGERATOR AND CHILL UNTIL FIRM.

Note:
Substitute pineapple juice for a portion of or all of the cold water. Pineapple can be drained in advance and juice chilled. Cottage cheese may be processed in a blender or food processor, if desired, before adding to the Jell-O.

Seven Layer Salad

Yield: 8 to 10 servings

INGREDIENTS:

LAYER THESE INGREDIENTS IN A GLASS BOWL OR RECTANGULAR DISH:

1 large head lettuce, torn

1-1/2 cups chopped celery

1/2 cup chopped green onion

1 cup diced green pepper

1 10 oz. pkg. frozen peas, thawed

3 or 4 hard-boiled eggs, chopped

1/2 pound hickory smoked bacon, cooked and coarsely chopped

2 cups mayonnaise

2 Tbsp. granulated sugar

grated Parmesan cheese

IN A LARGE GLASS BOWL or rectangular pan, begin with lettuce and add layers of celery, green onion, green pepper, peas, eggs and the bacon. "Frost" all ingredients with the mayonnaise and sprinkle on sugar.

TOP WITH PARMESAN CHEESE and cover with plastic wrap. Refrigerate overnight.

SPINACH SALAD

Yield: 4 salads

INGREDIENTS:

fresh spinach leaves, rinsed and stems removed

3 to 4 hard-boiled eggs, cut in wedges

1/2 pound hickory smoked bacon, cooked and chopped

red onion, sliced

ranch dressing

black olives

cherry tomatoes

seasoned croutons

ARRANGE SPINACH LEAVES (and other greens if desired) ON INDIVIDUAL SALAD PLATES. Add wedges of hard cooked egg, the bacon and the sliced red onion. Top with seasoned croutons and ranch dressing. Decorate salad plates with ripe olives and cherry tomatoes.

Waldorf Salad

Yield: 8 to 10 servings

INGREDIENTS:

6 apples, Red Delicious or Jonathan, not peeled

1 cup chopped celery

1 cup red or green seedless grapes, halved

1 cup chopped walnuts or pecans

1 cup miniature marshmallows (optional)

1 cup mayonnaise

1/2 cup sour cream

1 cup Cool Whip or whipped topping

1/2 tsp. ground nutmeg

CORE AND SLICE APPLES, REMOVING SEEDS. Cut them into small slices or chunks and place in a bowl. Add the celery, grapes, nuts and marshmallows.

MIX TOGETHER the mayonnaise, sour cream, Cool Whip or whipped topping and nutmeg. Combine with all other ingredients. Chill before serving.

Wilted Lettuce Salad

Yield: 2 to 4 servings

INGREDIENTS:

fresh salad greens

4 slices hickory smoked bacon, chopped

1 Tbsp. butter or margarine

1/4 cup apple cider vinegar

1 Tbsp. granulated sugar

salt and pepper to taste

paprika to taste

1/2 cup diced red onion

FRY BACON IN A SKILLET until crisp. Add to the bacon and drippings the butter or margarine, vinegar and sugar. Simmer until hot.

POUR HOT DRESSING over salad greens in a warm bowl or on individual warmed plates. Season with salt, pepper and paprika to taste. Top with diced red onion. Serve immediately.

Meats
The Main Entrée

Because I married into the meat business, I soon learned more about various cuts of meat and I researched different ways to prepare them. I borrowed ideas and recipes from family members, remembering the meals shared when I was growing up. My Czech grandfather was famous for his roast chicken or duck dinners, with dumplings and sauerkraut on the side. When my parents hosted Thanksgiving Dinner—of course, there was turkey and all the trimmings, but there was also a beautiful table setting for guests to enjoy, with china, glassware, and a linen tablecloth beneath it all.

Meats were, as a rule, the main entrée in our home. My husband, Frank, enjoyed pork cheeks smothered in gravy, chicken paprika, pot roast and steak on the grill, to name just a few. Potato dishes, vegetables and salads were a part of the mix, but meats were the mainstay.

My collection of meat recipes grew over a period of several years, as our customers asked for suggestions to prepare certain meals—therefore the ideas for meat entrées outnumber other recipes in this book.

Crock Pot Beef Chuck Roast

Yield: 4 to 6 servings

INGREDIENTS:

1 3 to 4 pound beef chuck roast

2 to 3 Tbsp. margarine

salt, black pepper and garlic powder to taste

1 pound carrots, peeled and cut in chunks

4 or 5 potatoes, peeled and halved or quartered

1 medium onion, coarsely chopped

1 pkg. onion gravy mix

1 14 oz. can diced tomatoes

SEASON BEEF ROAST AND BROWN ON BOTH SIDES in margarine in a skillet on top of the stove.

LAYER CARROTS AND POTATOES IN CROCKPOT. Place browned roast on top of these. Add the chopped onion and onion gravy mix. Pour the diced tomatoes over all and cook on LOW for about 6 to 7 hours, or on HIGH for about 4 to 5 hours. Meat should be very tender and vegetables will be done.

TRANSFER ROAST, POTATOES AND CARROTS TO A PLATTER OR COVERED DISH. Thicken the broth, if desired, with a little flour or cornstarch and spoon over meat and vegetables on serving platter.

Beef Pot Roast (Top of the Stove)

Yield: about 6 servings

INGREDIENTS:

1 3 to 3 1/2 pound beef chuck roast

2 to 3 Tbsp. butter or margarine

1 or 2 medium onions, quartered

1 pound carrots

4 to 6 medium potatoes

salt, pepper and garlic powder to taste

1 pkg. onion gravy mix

SEASON ROAST WITH SALT, PEPPER AND GARLIC POWDER. Brown in heavy deep skillet or Dutch oven on top of the stove (medium-high) in butter or margarine. Add onions and a little water. Shake on the onion gravy mix— then cover pan and simmer on LOW for approximately 2 to 3 hours, or until tender. Turn roast periodically and add small amounts of water as needed.

MEANWHILE, PREPARE CARROTS AND POTATOES:

Peel carrots and cut in halves or thirds diagonally. Place in a saucepan and cover with water.

Peel potatoes and cut in half. Place in a second saucepan and cover with water.

Cook the vegetables ONLY UNTIL NEARLY TENDER.

DRAIN WATER FROM THE COOKED CARROTS AND POTATOES.

Remove roast from skillet or Dutch oven and keep warm on a platter.

Place drained carrots and potatoes into the skillet or Dutch oven and allow to cook, turning to brown in the pan juices. A small amount of cornstarch dissolved in a little cold water may be added to thicken the meat juices.

To serve, arrange vegetables on the platter with the roast, and pour the thickened meat juices over all.

MARILYN STOYSICH

Beef Roast-Marinade

Yield: 6 to 8 servings

INGREDIENTS:

1 3 to 4 pound boneless beef roast

MARINADE:

1/2 cup dry red wine

1/2 cup tomato catsup

1/2 cup vegetable oil

1/2 cup Worcestershire sauce

1 clove garlic, minced

1 tsp. mustard seed

1 Tbsp. sweet basil leaves (or fresh basil)

1/2 tsp. coarse ground black pepper

1 tsp. salt

PREPARE MARINADE, COMBINING ALL INGREDIENTS IN ORDER GIVEN. Place beef roast in a deep bowl or a large plastic bag in bowl or glass dish—pour on the marinade. Cover bowl or twist bag closed.

MARINATE IN REFRIGERATOR several hours (or overnight). Turn roast once or twice to distribute marinade on all surfaces.

REMOVE ROAST FROM MARINADE and cook in roaster pan in a 325° oven until internal temperature of meat registers about 142° to 144° on meat thermometer.

Allow roast to rest a few minutes before carving.

Beef Short Ribs (Crock Pot Recipe)

Yield: 4 servings

INGREDIENTS:

2-1/2 pounds boneless OR bone-in beef short ribs

3 Tbsp. butter or margarine

black pepper to taste

garlic powder to taste

1 pkg. onion gravy mix

BROWN SHORT RIBS in butter or margarine in a skillet on top of the stove.

Transfer to crock pot and add black pepper, garlic powder and onion gravy mix. Add a small amount of water, and cook on LOW for about 4 hours, or until short ribs are tender. A small amount of cornstarch dissolved in a little cold water may be added to thicken the meat juices. Serve with noodles or mashed potatoes.

Braised Oxtails (Crock Pot Recipe)

Yield: 3 to 4 servings

INGREDIENTS:

2-1/2 pounds well-trimmed oxtails

3 Tbsp. butter or margarine

black pepper to taste

garlic powder to taste

1 pkg. onion gravy mix

BROWN OXTAILS in butter or margarine in a skillet on top of the stove.

Transfer to crock pot and add black pepper, garlic powder and onion gravy mix. Add a small amount of water, and cook on LOW for about 4 hours, or until oxtails are tender. A small amount of cornstarch dissolved in a little cold water may be added to thicken the meat juices. Serve with noodles or mashed potatoes.

BEEF STROGANOFF

Yield: 6 to 8 servings

INGREDIENTS:

1-1/2 pounds sirloin steak, cut in strips

1 tsp. garlic powder

1 tsp. coarse ground black pepper

1/2 cup flour

3 Tbsp. butter or margarine

1 cup water

1 pkg. beef stroganoff mix

2 Tbsp. butter or magarine

8 oz. fresh mushrooms, sliced

1 large onion, coarsely chopped

1 large green pepper, cut in julienne strips

1 10 oz. can cream of mushroom soup

3/4 cup sour cream

1 16oz. pkg. kluski noodles, cooked and drained

FLOUR AND SEASON THE SIRLOIN STEAK AND BROWN in a skillet with 3 Tbsp. butter or margarine over medium-high heat. Add water to the skillet and simmer on low for about 15 to 20 minutes. Add stroganoff mix and stir. Meanwhile, in another skillet, sauté mushrooms in 2 Tbsp. butter or margarine and add these to the steak. Sauté and add the chopped onion and green pepper.

ADD THE MUSHROOM SOUP TO SKILLET and continue cooking for another 20 minutes or so. Finally, stir in the sour cream. Serve Beef Stroganoff over hot, cooked and buttered noodles.

BEEF SWEETBREADS

Yield: about 3 to 4 servings

Beef sweetbreads come from the thyroid gland of the animal near the top of the chest. When properly cooked, they are tender and have a delicate flavor. The word 'sweetbread' came about because of the 'sweet' flavor, and also from the word 'Brede' or 'roasted meat.'

INGREDIENTS:

2 pounds beef sweetbreads

1/2 cup flour

2 eggs, beaten

1/2 cup milk

1 cup bread crumbs

salt and pepper to taste

4 to 6 Tbsp. margarine

COOK SWEETBREADS IN BOILING WATER for about 30 to 40 minutes. Remove from water, drain and cool.

WHEN COOL ENOUGH TO HANDLE, remove most of the membrane, being careful not to break up the sweetbreads if possible.

IN A SMALL BOWL, mix beaten eggs with milk.

DIP SWEET BREADS IN FLOUR, then in egg and milk batter, and finally in the bread crumbs. Allow to dry a few minutes on a plate dusted with flour. In a skillet melt the margarine and cook the sweetbreads over medium heat until golden brown.

Chicken A la King

Yield: about 6 servings

INGREDIENTS:

2-1/2 to 3 cups cooked chicken, cubed

2 cups fresh mushrooms, sliced

3 Tbsp. butter or margarine

1 cup chopped onion

1 large green pepper, seeded and chopped

1 15 oz. can chicken broth

3 Tbsp. cornstarch

2 10 oz. cans cream of mushroom soup

1 cup half & half or heavy cream

2 4 oz. jars diced pimento, drained

1 tsp. coarse ground black pepper

1 tsp. celery salt

LIGHTLY SAUTÉ mushrooms in butter or margarine in large skillet. Remove mushrooms and set aside.

IN SAME SKILLET, sauté just until tender the onion and green pepper. Remove and set aside.

MIX TOGETHER THE chicken broth and cornstarch and add to the skillet along with the mushroom soup and half & half or cream.

NEXT ADD CHICKEN, mushrooms, onion, green pepper, pimento, black pepper and celery salt.

ALLOW ALL INGREDIENTS TO THICKEN OVER LOW HEAT.

Chicken A la King is traditionally served over freshly baked biscuits. May also be served over cooked noodles or rice.

Lemon Chicken

Yield: 4 servings

INGREDIENTS:

1 chicken, quartered

1/2 cup butter or margarine

1/4 cup lemon juice

lemon pepper seasoning to taste

1/4 tsp. dill weed

salt and pepper to taste

PLACE CHICKEN QUARTERS IN BAKING PAN OR ROASTER. Broil a few minutes to brown. Add a small amount of water to the pan and turn oven control to Bake at 325°. Melt butter or margarine in a saucepan, add lemon juice, and begin basting chicken quarters.

SHAKE ON SEASONINGS, continue basting, and bake until a meat thermometer registers 165° to 170° when inserted into breast portion of the chicken.

BAKED CHICKEN WITH STUFFING

Yield: about 6 servings

INGREDIENTS:

1 3 to 4 pound chicken

1/2 cup chopped onion

1 cup chopped celery

1/2 cup margarine

3 cups cubed dry bread

1 15 oz. can chicken broth

salt and pepper to taste

1/2 tsp. poultry seasoning

black pepper to taste

1 Tbsp. parsley flakes

SAUTÉ ONION AND CELERY in 3 tablespoons of the margarine for about 3 to 5 minutes.

COMBINE IN A BOWL the cubed bread, the cooked onion and celery (margarine also), chicken broth, salt and pepper to taste and 1/4 tsp. of poultry seasoning. Dressing should be moist but not too wet. Spoon dressing into cavity of the chicken and secure with wooden picks or skewers.

PLACE CHICKEN IN A ROASTER PAN and coat with remaining margarine mixed with 1/4 tsp. poultry seasoning and black pepper and the parsley flakes.

BAKE IN A 350° OVEN for about 30 minutes. Reduce oven temperature to 325°. Baste occasionally with pan juices, adding a little water to the pan if needed. Continue roasting for another 30 minutes or so, or until internal temperature of chicken registers 165° to 170° on meat thermometer.

CHICKEN CACCIATORE

Yield: 4 to 6 servings

INGREDIENTS:

3 whole bone-in chicken breasts, split

1/2 to 3/4 cup flour

1/4 cup vegetable oil

1 cup water

2 14 oz. cans stewed tomatoes

2 cups chopped celery

2 large onions, halved and sliced

1 cup sliced zucchini

2 cups mushrooms, sliced and sautéed in 2 Tbsp. butter or margarine

1 Tbsp. Italian seasoning

1 tsp. ground oregano

2 tsp. garlic powder

salt and pepper to taste

COAT CHICKEN BREASTS WITH FLOUR and brown in a large skillet in vegetable oil over medium-high heat. Discard excess oil and add water. Simmer on low for about 20 minutes, turning chicken once or twice.

ADD STEWED TOMATOES, CELERY, ONION, ZUCCHINI, MUSHROOMS and all seasonings. Continue simmering over low heat for 20 to 30 minutes, or until chicken is tender. A small amount of cornstarch mixed with cold water may be used to thicken the sauce.

Serve with fettuccine or angel hair pasta.

Chicken and Noodles

Yield: 4 to 6 servings

INGREDIENTS:

1 cut-up chicken

3/4 cup flour

1/4 cup vegetable oil

1 cup water

1/2 tsp. coarse ground black pepper

1/2 tsp. celery salt

1 small onion, chopped

1 cup chopped celery

1 cup sliced carrots

1 10 oz. can cream of mushroom soup

1 10 oz. can cream of chicken soup

1 12 oz. pkg frozen noodles, cooked

FLOUR CHICKEN AND FRY IN VEGETABLE OIL IN A SKILLET ON TOP OF THE STOVE. When browned, eliminate some of the excess fat and add water. Add seasonings and begin simmering over low heat.

AFTER ABOUT 20 MINUTES, ADD THE VEGETABLES and more water if needed. Continue simmering on low (occasionally move around the chicken pieces) for about 30 minutes more or until a meat thermometer registers 165° when inserted into a chicken breast.

AT THIS TIME, REMOVE CHICKEN PIECES FROM THE SKILLET, and add both cans of soup. Stir so that soup forms the gravy, then return chicken to skillet (remove chicken from bone if preferred).

SERVE OVER COOKED, BUTTERED NOODLES.

VARIATION:
For Chicken and Dumplings, prepare recipe as above, and serve with Easy Potato Dumplings, found on page 144 of the "Potatoes" Section.

Oven BBQ Chicken

Yield: 4 servings

INGREDIENTS:

1 chicken, quartered

1 18 oz. bottle BBQ sauce, any brand

PLACE CHICKEN IN BAKING PAN OR ROASTER. Broil a few minutes to brown.

TURN OVEN CONTROL TO BAKE at 325° and baste chicken pieces with barbeque sauce at intervals until thoroughly cooked and internal temperature of 165° is reached.

Chicken Paprikás (Paprika Chicken)

Yield: 4 to 6 servings

INGREDIENTS:

1 cut-up chicken

Hungarian paprika

1/4 cup vegetable oil

1 cup water

1 large onion, coarsely chopped

salt and pepper to taste

1 cup or more sour cream

12 oz. frozen egg noodles, cooked

SHAKE A GENEROUS AMOUNT OF PAPRIKA ON CHICKEN PIECES, as if dusting with flour. Sauté chicken in a skillet with vegetable oil over medium heat, turning often to brown on all sides. Remove excess oil from skillet.

ADD WATER TO THE SKILLET AND LOWER HEAT. Add onion, salt and pepper and simmer for about an hour, occasionally turning chicken pieces—add more water if needed.

REMOVE THE CHICKEN AND KEEP WARM. Thicken pan juices with flour or cornstarch and stir in the sour cream. Return chicken to the skillet and heat through.

SERVE OVER COOKED BUTTERED NOODLES and shake on additional paprika if desired.

CHICKEN PARMESAN

Yield: 4 to 6 servings

INGREDIENTS:

6 4 oz. boneless chicken breasts

1 cup flour

1 egg mixed with 1/2 cup milk

1 cup seasoned bread crumbs

1/4 cup vegetable oil or margarine

coarse ground black pepper to taste

Italian seasoning to taste

1-1/2 cups spaghetti sauce or marinara sauce

1 cup shredded mozzarella cheese

COAT CHICKEN BREASTS WITH FLOUR, dip in egg and milk mixture, then in seasoned bread crumbs. Sauté in cooking oil or margarine over low heat in a skillet on top of the stove.

Place chicken breasts in a baking dish with spaghetti or marinara sauce, and bake in a 325° oven for about 20 to 30 minutes. Top with shredded cheese and return to oven (uncovered) until cheese melts.

SERVE WITH PASTA AND ADDITIONAL SAUCE.

Chili Con Carne

Yield: about 10 servings

INGREDIENTS:

3 pounds ground beef

1 cup chopped onion

2 cups finely chopped celery

1 or 2 large green peppers, chopped

2 pkgs. chili seasoning

salt and pepper to taste

1 tsp. garlic powder

1 Tbsp. ground cumin

2 14 oz. cans diced tomatoes

1 12 oz. can tomato juice

2 10 oz. cans tomato puree

1 15 oz. can dark red kidney beans

1 15 oz. can seasoned chili beans

1 15 oz. can red beans

OPTIONAL TOPPINGS:

cheddar cheese, shredded

sour cream

chopped onion

BROWN AND CRUMBLE GROUND BEEF. Add chopped onion, celery and green pepper. Continue cooking a few minutes. Drain fat and add chili seasoning packets and all other seasonings. Add tomatoes, tomato juice and tomato puree and simmer on low for approximately 20 to 30 minutes. Finally, add canned beans, and more water or tomato juice if chili mixture is too thick. Continue cooking on lowest heat until flavors blend—about 10 to 15 additional minutes.

SERVE WITH SALTINE CRACKERS and toppings of chopped onion, shredded cheddar cheese and sour cream.

Chorizo Sausage and Rice

Yield: 4 to 6 servings

INGREDIENTS:

1 pound fresh chorizo sausage

2 medium onions, sliced

1 green pepper, sliced into rings

4 cups cooked rice

2 Tbsp. butter

1 clove garlic minced

salt and pepper to taste

1 tsp. ground cumin

1 tsp. paprika

1 15 oz. can diced tomatoes

1 15 oz. can tomato sauce

1 cup frozen peas, thawed

REMOVE SAUSAGE FROM CASINGS. Brown and crumble in large skillet until nearly done. Add onions and green pepper rings and continue to cook over medium heat for 10 to 15 minutes.

Add rice, butter, garlic, seasonings and all other ingredients.

COVER AND SIMMER about 15 to 20 minutes. Stir once or twice while cooking.

Add fresh chopped cilantro if desired.

CORN DOGS

Yield: about 8 corn dogs

INGREDIENTS:

1 pound natural casing hot dogs

1 cup all-purpose flour

1-1/2 tsp. baking powder

1/2 tsp. salt

2 Tbsp. cornmeal

3 Tbsp. shortening

1 beaten egg

3/4 cup milk

wooden skewers

HEAT COOKING OIL IN DEEP-FAT FRYER TO 360°. Measure flour, baking powder, salt and cornmeal into a bowl. Cut in shortening, then stir in egg and milk.

DIP HOT DOG INTO BATTER, allowing excess batter to drip into bowl. (Batter should be thick enough to stick to the hot dogs.)

FRY CORN DOGS IN THE HOT OIL, turning once, until brown. Drain on paper towels and carefully insert wooden skewers into ends of the hot corn dogs to serve.

Corned Beef and Cabbage

Yield: 8 to 10 servings

Although Corned Beef and Cabbage is the standard fare for many Americans on St. Patrick's Day, in Ireland a hearty stew made with lamb is often served. The term "Corned Beef" came about because beef was traditionally stored in barrels with coarse grains or 'corns' of salt.

INGREDIENTS:

4 to 6 pound corned beef brisket

2 medium heads cabbage, cut into wedges

4 potatoes, peeled and halved

6 carrots, peeled and thickly sliced

2 onions, quartered

salt and pepper to taste

PLACE CORNED BEEF IN LARGE KETTLE OR DUTCH OVEN, and add water to cover. Slowly bring to a boil. Lower heat and cover kettle. Simmer approximately 2-1/2 to 3 hours or until beef is tender.

REMOVE CORNED BEEF AND KEEP WARM in a covered pan in the oven. To the cooking liquid add the potatoes, carrots, onions and salt and pepper as desired. Add water if needed. Increase heat, cover and cook for 10 minutes. Add cabbage and cook an additional 10 to 15 minutes. Remove vegetables from kettle and combine with Corned Beef to serve— Meat should be cut against the grain.

Roast Duckling

Yield: 4 servings

INGREDIENTS:

1 domestic duckling
salt and pepper to taste
poultry seasoning

THAW DUCKLING (if frozen) ACCORDING TO PACKAGE DIRECTIONS. Remove plastic wrapper and rinse with cold water. Wipe cavity with paper towel, trim away excess fat and salt the cavity. Stuff with Wild Rice Dressing (on page 160 of the "Rice" section) and sew up opening. Neck cavity should also be stuffed.

PLACE DUCKLING BREAST SIDE UP ON A RACK in a large roasting pan. Season with salt and pepper and a little poultry seasoning. Bake in a 325° oven approximately 2-1/2 to 3 hours. Add water to bottom of pan as needed to prevent spattering of grease—and/or remove some of the drippings from pan. Roaster lid may be used at start of baking time—then removed to allow the duckling to brown well. When done, internal temperature should register about 170° on meat thermometer.

VARIATION: DUCK A L'ORANGE

ORANGE SAUCE:
1 cup orange marmalade
1/2 cup chicken stock
1 Tbsp. cornstarch

HEAT ORANGE SAUCE INGREDIENTS in a saucepan on top of stove until slightly thickened.

PREPARE AND BAKE DUCKLING PER RECIPE ABOVE.

During last 20 minutes of baking time, baste with orange sauce.

Easy Oven Dinner

Yield: 6 to 8 servings

INGREDIENTS:

- 1 3 to 4 pound beef chuck roast
- salt, pepper and garlic powder to taste
- 6 medium potatoes
- 1 pound carrots
- boiling water
- 2 medium onions
- 1 8 oz. pkg. whole mushrooms
- 1 15 oz. can diced or stewed tomatoes
- 2 pkgs. onion gravy mix or brown gravy mix
- 3 Tbsp. flour
- 1-1/2 cups water

PEEL POTATOES, cut in halves. Peel carrots, and cut diagonally into two to three inch pieces, depending on size of carrots. Peel onions and quarter. Place potatoes and carrots in large Pyrex dish or bowl. Pour on boiling water to cover—let stand 10 minutes and drain.

SEASON ROAST WITH SALT, BLACK PEPPER AND GARLIC POWDER.

ARRANGE ROAST AND ALL VEGETABLES in large oval roaster or deep roasting pan.

DISTRIBUTE CONTENTS OF ONE PACKAGE OF GRAVY MIX OVER THE ROAST AND VEGETABLES. Mix second package with 3 tablespoons flour and 1-1/2 cups water in measuring cup. Mix well. Pour evenly over just the vegetables.

COVER WITH ROASTER LID OR HEAVY FOIL. Bake in a 300° oven for approximately 3 1/2 to 4 hours.

Hamburger Goulash

Yield: 8 to 10 servings

INGREDIENTS:

2 pounds ground beef

1 cup chopped onion

1 cup diced green pepper

salt and coarse ground black pepper to taste

1 tsp. garlic powder

1/2 tsp. ground cumin

1 tsp. chili powder

16 oz. shell or elbow macaroni

1 14 oz. can stewed tomatoes

1 26 oz. can or jar spaghetti sauce

BROWN AND CRUMBLE the ground beef. Add the onion and green pepper and cook 10 to 15 minutes. Add salt, pepper, garlic powder, cumin and chili powder.

IN A LARGE KETTLE BRING WATER TO A BOIL. Cook macaroni according to package directions, drain and rinse. Return macaroni to kettle. Add the cooked, seasoned ground beef, onion and green pepper, draining excess fat if needed.

ADD THE STEWED TOMATOES AND SPAGHETTI SAUCE. Simmer on low a few more minutes and serve.

Sausage and Chicken Gumbo

Yield: 6 to 8 servings

INGREDIENTS:

2 8 oz. boneless chicken breasts

1/2 cup flour

1/3 cup vegetable oil

4 cups water

2 14 oz. cans stewed tomatoes

1 large green pepper, seeded and chopped

1 large onion, chopped

1/2 tsp. salt

1 tsp. coarse ground black pepper

2 cups sliced okra, fresh or frozen

1 pound smoked Polish or smoked Cajun sausage, sliced

1 tsp. thyme leaves

2 cloves finely minced garlic

1 Tbsp. granulated sugar

1/4 to 1/2 tsp. crushed red pepper

1 or 2 bay leaves

3 tsp. Filé Powder

6 to 8 cups hot cooked rice

COAT THE CHICKEN WITH FLOUR AND BROWN THE CHICKEN in vegetable oil in a kettle over medium heat. Add the water and maintain a slow simmering boil. When chicken registers 160° on meat thermometer, remove from kettle and cool. Cut up chicken and return to kettle.

ADD: stewed tomatoes, green pepper, onion, salt, black pepper, okra, sausage, thyme leaves, garlic, sugar, crushed red pepper and bay leaves. Simmer on low for 45 minutes.

Add the Filé Powder and simmer for 10 to 15 more minutes.

REMOVE BAY LEAVES before serving in bowls over hot cooked rice.

Cherry Glazed Ham

Yield: 12 or more servings

INGREDIENTS:

1 hickory smoked bone-in ham

whole cloves

1 12 oz. jar cherry preserves

1 cup brown sugar

2 tsp. prepared mustard

1 cup red wine

PLACE HAM IN ROASTER PAN. Score top of ham in a diamond pattern, and stud with whole cloves. Place in a 325° oven and bake approximately 1-1/2 to 2 hours, or until internal temperature registers 140° on meat thermometer.

WHILE HAM IS BAKING, combine cherry preserves, brown sugar, prepared mustard and red wine in a saucepan on top of the stove. Simmer a few minutes, and then allow to cool.

DURING THE LAST 30 MINUTES OF BAKING TIME, spoon glaze over the ham at frequent intervals.

ORANGE-PINEAPPLE GLAZE:

1 15 oz. can crushed pineapple

1 Tbsp. vegetable oil

1 cup orange marmalade

1/2 cup brown sugar

VARIATION: ORANGE-PINEAPPLE GLAZE

BRING ALL INGREDIENTS TO A BOIL IN SAUCEPAN. Remove from heat and cool slightly. Spoon glaze over ham several times during the last 30 minutes of baking time. This glaze may also be used for roast duckling.

KNOCKWURST AND KRAUT

Yield: 6 to 8 servings

INGREDIENTS:

1-1/2 pounds knockwurst sausage

2 Tbsp. margarine

1 onion, quartered and sliced

1 27 oz. can sauerkraut, drained

1 14 oz. can whole potatoes, drained

3 Tbsp. brown sugar

3 Tbsp. margarine

1 tsp. caraway seed

salt and pepper to taste

1 cup water

MELT MARGARINE IN A SKILLET and lightly sauté onion. Place sauerkraut in a casserole dish. Top sauerkraut with the sautéed onion, potatoes and brown sugar. Next add the margarine, caraway seed and salt and pepper to taste. Pour on the water and arrange sausages on top.

BAKE, COVERED, in a 325° oven for about 45 minutes to one hour. Remove casserole lid to turn sausage, and bake an additional 15 minutes.

LEG OF LAMB

Yield: 8 to 10 servings

INGREDIENTS:

1 5 pound boneless leg of lamb

2 cloves fresh garlic, slivered

BASTING SAUCE:

1 cup melted butter

1 tsp. coarse ground pepper

1 Tbsp. minced garlic

1 Tbsp. parsley flakes

1 tsp. salt

2 Tbsp. Worcestershire sauce

PLACE LEG OF LAMB FAT SIDE UP IN A ROASTING PAN. Insert slivers of fresh garlic under the skin of the lamb using a pointed knife. Bake, covered, in a 350° oven for about 30 minutes. Lower oven temperature to 325°. Continue roasting for a total of 2-1/2 to 3 hours, or until internal temperature reaches 160° on a meat thermometer. During the last hour of roasting time, remove lid, and baste frequently with sauce.

Slow-Cooked Lamb Shanks

Yield: 4 servings

INGREDIENTS:

4 fresh lamb shanks, whole

salt and pepper to taste

1/2 tsp. garlic powder

1/2 cup flour

1/2 cup vegetable oil

1 medium onion, chopped

4 russet potatoes, unpeeled and cut in half, lengthwise

1 pound carrots, peeled and cut in chunks

1 Tbsp. fresh rosemary, minced

1 pkg. onion gravy mix

1-1/2 cups water

dash Worcestershire sauce

SEASON LAMB SHANKS, COAT WITH FLOUR AND BROWN on all sides in oil in a skillet on top of the stove. Remove from skillet and place in crock pot.

NEXT ADD the chopped onion, potatoes (cut side down) and the carrots. Add fresh rosemary and the onion gravy mix. Pour in water and a dash of Worcestershire sauce. Cook on LOW about 6 hours. A small amount of cornstarch dissolved in a little cold water may be added to thicken the meat juices.

LIVER AND ONIONS

YIELD: 4 to 6 servings

INGREDIENTS:

4 to 6 slices fresh beef liver

flour (about 2 Tbsp.)

salt and pepper to taste

1/4 tsp. garlic powder

3 Tbsp. margarine

1 cup water

1 large onion, sliced

1 pkg. brown gravy mix

COAT LIVER WITH FLOUR, ADDING SALT, PEPPER AND GARLIC POWDER. Sauté the liver in margarine on medium heat until browned on both sides. Add water, sliced onion and the brown gravy mix.

SIMMER SLOWLY UNTIL LIVER IS TENDER, turning as needed, and adding water if necessary. A small amount of cornstarch dissolved in a little cold water may be added to thicken the meat juices.

Serve with mashed potatoes.

Mexican Lasagna

Yield: 10 to 12 servings

INGREDIENTS:

2 pounds ground beef

1/2 cup diced onion

1 green pepper, chopped

1 tsp. garlic, minced

salt and pepper to taste

1 pkg. taco seasoning mix

1 10 oz. can enchilada sauce

1 14 oz. can diced tomatoes

1 15 oz. can tomato sauce

10 corn tortillas

vegetable oil

2 cups ricotta cheese

2 eggs, beaten

1/2 pound shredded Swiss-American cheese

1/2 pound shredded cheddar cheese

1 4 oz. can chopped black olives

tortilla chips, crushed

BROWN AND CRUMBLE GROUND BEEF in a skillet. Drain fat and return to pan. Add onion, green pepper, garlic and seasonings, stirring over medium heat. Add enchilada sauce, diced tomatoes, tomato sauce and lower heat. FRY CORN TORTILLAS in hot oil just enough to soften, and drain on paper towels. When cool, cut them into halves.

Mix ricotta cheese with eggs and assemble casserole as follows:

INTO A 9x13 PAN, spoon 1/3 of meat and tomato mixture and top with 1/2 the Swiss-American cheese. Next, spoon on half of the ricotta cheese-egg mixture and spread evenly. Arrange 5 tortilla shells (10 halves) over this and then add another 1/3 of the meat-tomato mixture. Now add the rest of the Swiss-American cheese and remaining ricotta cheese. Arrange the remaining tortilla halves on top of this, and finally the rest of the meat and tomato mixture. Top all with the cheddar cheese, black olives, and crushed tortilla chips. Bake in a 325° oven for approximately 25 to 30 minutes. This dish may be put together in advance and refrigerated—allow additional time in the oven, or bring to room temperature before baking.

Italian Meatballs

Yield: 6 to 8 servings

INGREDIENTS:

2 pounds ground beef

1/2 pound bulk pork sausage

1 medium onion, minced

1 cup seasoned bread crumbs

1 tsp. salt

1 tsp. coarse ground black pepper

1/2 tsp. garlic powder

1/3 cup grated Parmesan cheese

2 eggs, beaten

dash of Italian seasoning

water to moisten

MIX ALL INGREDIENTS AND FORM INTO ONE OR TWO INCH BALLS. Bake on a greased cookie sheet in a 350° oven for approximately 30 minutes or until meatballs are thoroughly cooked. Turn once or twice during baking time. Heat in spaghetti sauce in the oven or on top of the stove.

Serve with pasta.

SERVING SUGGESTION: MEATBALL SANDWICHES

TOAST SPLIT HOAGIE BUNS AND HOLLOW OUT A PORTION OF THE LOWER HALF OF BUN. Add cooked meatballs, spaghetti sauce or marinara sauce and shredded mozzarella cheese. Place under broiler until cheese melts and add the top portion of bun.

Old-Fashioned Meatloaf

Yield: 6 to 8 servings

INGREDIENTS:

2 pounds ground beef

1/2 pound bulk pork sausage

1/2 cup finely chopped onion

1/2 cup BBQ sauce or Heinz 57 Sauce, as preferred

2 eggs, beaten

1/2 cup water

1 cup seasoned bread crumbs

1/2 tsp. Lawry's seasoned salt

1 tsp. coarse ground black pepper

1 tsp. garlic powder

BBQ sauce for topping

COMBINE GROUND BEEF, PORK SAUSAGE AND ALL REMAINING INGREDIENTS. Form mixture into a loaf in a baking pan. Bake in a 325° to 350° oven for about 30 minutes.

COVER WITH FOIL AND CONTINUE BAKING for at least 30 more minutes. Uncover, coat meat loaf with one cup or more of BBQ sauce and return to the oven for 15 more minutes. Internal temperature of meat loaf should register 160° to 165° on meat thermometer.

Mexican Pork Stew (Posole)

Yield: 4 to 6 servings

INGREDIENTS:

1-1/2 pounds pork tenderloin, cut into 1 inch cubes

4 Tbsp. margarine

1 medium onion, diced

3 cloves garlic, chopped fine or minced

2 15 oz. cans white hominy

1 4 oz. can chopped green chiles

2 15 oz. cans chicken broth

1 10 oz. can tomato puree

1 tsp. ground cumin

1/2 tsp. ground oregano

3 potatoes, peeled and cut into chunks

sour cream

cilantro

MELT HALF THE MARGARINE IN LARGE HEAVY SKILLET and sauté onion and garlic until transparent. Remove from pan and set aside.

MELT REMAINING 2 TABLESPOONS of margarine and brown pork over medium high heat, searing meat well. Lower temperature and add hominy, green chiles, the chicken broth, tomato puree and the seasonings. Simmer on low to medium heat for about 20 minutes. Add the sautéed onions, garlic, and potatoes to the pan, and continue cooking approximately 15 to 20 more minutes, or until potatoes are done and meat is tender.

SERVE IN SOUP BOWLS TOPPED WITH A DOLLOP OF SOUR CREAM and a sprinkling of chopped fresh cilantro. Shake on a small amount of crushed red pepper if desired.

Stuffed Peppers

Yield: 6 or more servings

INGREDIENTS:

1-1/2 pounds ground beef

1/2 pound bulk pork sausage

1-1/2 cups uncooked rice

1 egg, beaten

1/2 tsp. salt

1/2 tsp. garlic powder

1 tsp. coarse ground black pepper

1 to 1-1/2 cups water

6 or 7 green bell peppers

2 14 oz. cans diced or stewed tomatoes

2 tsp. granulated sugar

1 8 oz. can tomato sauce

1/2 tsp. Kitchen Bouquet

IN A BOWL, MIX BEEF, PORK, RICE, EGG AND SEASONINGS, adding enough water to maintain a medium-soft consistency.

WASH, CUT AWAY STEMS AND REMOVE SEEDS from the green bell peppers.

TO KETTLE OR STOCKPOT, add one can of the tomatoes and 2 teaspoons sugar. (A little sugar helps cut the bitterness of the green peppers.)

STUFF PEPPERS WITH FILLING and put these in the kettle. Top with the second can of tomatoes and the tomato sauce.

COOK ON TOP OF THE STOVE ON LOW HEAT, for 1 to 1-1/2 hours. Move peppers around with a wooden spoon occasionally to prevent sticking to the bottom of kettle. At the end of cooking time, extract some of the broth, thicken slightly with flour or cornstarch, stir in the Kitchen Bouquet, and return mixture back to the ketttle for 5 minutes.

Serve with mashed potatoes.

Stuffed Peppers-Italian Style

Yield: 12 to 14 pepper halves

INGREDIENTS:

6 or 7 large green bell peppers

1-1/2 pounds ground beef

1/4 pound bulk pork sausage

1 egg, beaten

1 small onion, chopped

3/4 cup seasoned bread crumbs

1 cup uncooked white rice

1/4 cup grated Parmesan cheese

1 tsp. Italian seasoning

1/2 tsp. garlic powder

1 tsp. coarse ground black pepper

1/2 tsp. salt

water

1 14 oz. can diced tomatoes

2 tsp. granulated sugar

1 26 oz. jar pasta or marinara sauce

2 cups shredded Italian cheese blend

CUT CAREFULLY AROUND STEM and remove tops of peppers. Cut each in half and remove seeds and membrane - Rinse halves and set aside.

IN LARGE BOWL mix ground beef, pork sausage, egg, onion, bread crumbs, rice, Parmesan cheese and seasonings. Add water until mixture is of medium consistency.

POUR DICED TOMATOES into a large baking pan, and add granulated sugar. (A little sugar helps cut the bitterness of the green peppers.)

STUFF PEPPER HALVES AND PLACE IN PAN. Cover pan with foil and bake in a 375° oven for about an hour. Remove foil, spoon on the pasta or marinara sauce and return to oven (uncovered) about 15 minutes. Top with shredded cheese and bake for a few more minutes just until cheese melts.

Pork Cheeks

Yield: 4 to 6 servings

INGREDIENTS:

2 pounds pork cheek meat

1/2 cup flour

salt, pepper and garlic powder to taste

1/2 cup vegetable oil

small amount of water

1 large onion, chopped

FLOUR AND SEASON PORK CHEEKS. Sauté them in hot oil until brown on all sides. Remove from pan and eliminate some of the fat.

RETURN PORK CHEEKS TO PAN, add water and onion and simmer on low for about two hours or until tender. Turn periodically, adding water if needed. Add a small amount of cornstarch dissolved in a little cold water to thicken the pan juices for gravy.

Serve with mashed potatoes.

Note: Pork cheeks are the facial muscle of the hog. Long slow cooking makes them very tender and delicious.

Pork Chops and Gravy

Yield: 6 servings

INGREDIENTS:

- 6 bone-in pork chops
- 1/2 cup milk
- 1 cup flour
- vegetable oil
- water
- salt and pepper to taste
- 1/2 tsp. garlic powder
- 1 medium onion, chopped
- 1 10 oz. can cream of mushroom soup
- 1 cup sour cream
- 1/2 tsp. Kitchen Bouquet

MOISTEN PORK CHOPS WITH MILK AND DREDGE WITH FLOUR. Brown with a little oil in a skillet on top of the stove. When pork chops are browned on both sides, eliminate some of the fat from skillet and add a little water. Season chops as desired and add onion. Simmer on low for about 45 minutes to one hour, turning occasionally and adding water if needed.

TRANSFER CHOPS TO PLATTER and add soup, sour cream and Kitchen Bouquet to skillet. Stir and cook mixture over low heat and return chops to pan. Continue simmering a few more minutes—chops should be very tender.

Serve Pork Chops and Gravy with mashed potatoes.

Stuffed Pork Chops

Yield: 6 servings

INGREDIENTS:

6 thick cut rib pork chops

4 Tbsp. margarine

1/2 cup finely chopped onion

1/2 cup finely chopped celery

2 cups dried or day-old bread, cut into small pieces

1 cup seasoned bread crumbs

1/4 tsp. poultry seasoning

salt and pepper to taste

1 cup chicken broth

1/4 tsp. coarse ground black pepper

1/4 tsp. garlic powder

WITH A SHARP KNIFE cut a pocket into each pork chop.

MELT MARGARINE IN A SKILLET and sauté onion and celery until nearly tender.

IN A BOWL mix the sautéed onion and celery, including pan drippings, with the cubed bread, bread crumbs, poultry seasoning, salt, pepper, and chicken broth.

SPOON ABOUT 2 Tbsp. of stuffing into the pocket of each pork chop. Secure with wooden picks, and season with black pepper and garlic powder.

PLACE CHOPS in a greased 9x13 baking pan. Cover with foil and bake in a 350° oven for about 45 minutes to one hour or until meat thermometer inserted into the pork chop reads 160°.

Pork Chow Mein

Yield: 10 to 12 servings

INGREDIENTS:

2 Tbsp. vegetable oil

2 Tbsp. margarine

3/4 cup chopped onion

1 cup chopped celery

2 cups fresh broccoli florets

1 cup fresh cauliflower florets

2 cups sliced fresh mushrooms

1 pound fresh bean sprouts

2 cups snow peas

garlic salt to taste

coarse ground black pepper to taste

1 tsp. Chinese 5-spice powder

4 cups cooked pork, cubed

2 8 oz. cans sliced water chestnuts, drained

1 4 oz. jar sliced pimento

2 Tbsp. cornstarch

1 cup chicken broth

IN HEAVY NON-STICK SKILLET OR WOK heat oil and margarine. Sauté fresh vegetables on medium-high until nearly tender—vegetables should be crisp, not over-cooked. Transfer with slotted spoon to 6-quart kettle or Dutch oven. Add small amounts of additional margarine to skillet and continue until all fresh vegetables are sautéed. Season while cooking with the garlic salt, black pepper and Chinese 5-spice powder.

HEAT THE PORK in a skillet to brown a little and enhance the flavor; then, add to the other ingredients in the kettle. Add water chestnuts and sliced pimento.

MIX CORNSTARCH WITH chicken broth. Add to ingredients in kettle, and allow to simmer a few more minutes.

Serve over hot cooked long-grain rice, or Chinese noodles.

Soy sauce and chopped green onion compliment this dish.

German Pork Cutlets

Yield: 6 servings

INGREDIENTS:

6 pork cutlets

1 cup flour

1/2 tsp. salt

1/2 tsp. black pepper

2 eggs, well beaten

1/2 cup milk

1 cup seasoned bread crumbs

2 tsp. paprika

1/2 cup vegetable oil

COMBINE FLOUR WITH SALT AND PEPPER. Coat pork cutlets on both sides with seasoned flour. Combine beaten eggs with milk, dip cutlets into the milk-egg mixture, and then in bread crumbs to which paprika has been added.

HEAT VEGETABLE OIL in a skillet and brown cutlets (medium heat) approximately 3 to 5 minutes on each side. Transfer the cutlets to a covered roasting pan and bake in a 325° oven for about 30 minutes.

Serve with hot buttered noodles and Sour Cream-Dill Gravy recipe on page 193 of "Sauce and Gravy" section.

Polish Style Pork Loin

Yield: 8 to 10 servings

INGREDIENTS:

1   3 to 4 pound pork loin roast
2 Tbsp. flour
1 tsp. granulated sugar
1/2 tsp. salt
1/2 tsp. black pepper
1 tsp. dry mustard
1/2 tsp. ground sage

TOPPING:
1 cup applesauce
1/2 cup brown sugar
1/2 tsp ground cinnamon
1/4 tsp. ground cloves
1/4 tsp. salt

COMBINE APPLESAUCE, BROWN SUGAR, cinnamon, cloves and salt. Set aside.

MIX FLOUR, SUGAR AND SEASONINGS and rub into the pork loin. Place pork in roaster pan and bake uncovered in a 325° oven for approximately 1-1/2 hours.

BEGIN SPOONING APPLE TOPPING over the pork loin at intervals, until all is used.

BAKE APPROXIMATELY ONE HOUR LONGER, OR UNTIL EVENLY BROWNED. Internal temperature of the pork loin should read 145° to 160° on meat thermometer. Allow meat to rest before serving.

PORK STIR-FRY

Yield: 6 to 8 servings

INGREDIENTS:

1-1/2 pounds boneless pork tenderloin, cut in strips

coarse ground black pepper to taste

5 Tbsp. margarine

1/2 tsp. garlic powder

1 tsp. Chinese 5-spice powder

2 cups peeled and sliced carrots

1-1/2 cups chopped celery

1 small onion, coarsely chopped

2 cups broccoli florets

1 10 oz. pkg. frozen sugar snap peas

1 red bell pepper, cut in strips

1 green bell pepper, cut in strips

1 cup sliced fresh mushrooms

1 8 oz. can sliced water chestnuts, drained

1/2 tsp. Kitchen Bouquet

6 to 8 cups cooked rice

1 bunch green onion, chopped

soy sauce to taste

SEASON THE PORK WITH BLACK PEPPER and sauté in 3 Tbsp. of margarine in a deep skillet on top of the stove. When browned, remove from skillet and set aside.

IN THE SAME SKILLET add 2 Tbsp. margarine and sauté vegetables, beginning with carrots, celery and onion. Add garlic powder, Chinese 5-spice powder, broccoli, sugar snap peas, and red and green pepper and cook until these vegetables are just tender. Last, add mushrooms and water chestnuts to the skillet and a little more margarine if needed

COMBINE THE COOKED PORK WITH VEGETABLES and continue to cook a few more minutes. Extract some pan juices and thicken slightly with a little cornstarch mixed first in cold water then added to the skillet, along with the Kitchen Bouquet. Simmer a few more minutes.

SERVE PORK STIR FRY over hot cooked rice.

Top with chopped green onion and soy sauce.

Pork and Sweet-Sour Cabbage

Yield: 6 to 8 servings

INGREDIENTS:

8 one-inch thick center cut pork chops

salt and pepper to taste

1/2 cup flour

1/2 cup vegetable oil

2 medium heads cabbage, cores removed and coarsely chopped

1 large onion, chopped

1 tsp. caraway seed

1/4 tsp. salt

3 Tbsp. flour

3 Tbsp. brown sugar

1-1/2 cups water

3 Tbsp. vinegar

HEAT OIL IN A HEAVY SKILLET.

Season and flour pork chops and brown on both sides.

Transfer to platter and keep warm.

PLACE CABBAGE AND ONION IN ROASTER PAN. Arrange browned pork chops on top. Mix together the caraway seed, salt, flour, brown sugar, water and vinegar. Pour over pork chops.

Cover roaster pan and bake in a 325° oven for 20 minutes. Lift the lid and gently stir cabbage away from the sides of the pan. Replace lid, return to oven and lower oven temperature to 275°—there should be sufficient liquid to cook the cabbage. Continue baking approximately 30 minutes. The pork chops should be thoroughly done at this time.

Reuben Casserole

Yield: about 8 servings

INGREDIENTS:

4 Tbsp. margarine

4 Tbsp. flour

3-1/2 cups milk

2 Tbsp. Dijon mustard

6 potatoes, peeled and sliced

1 27 oz. can sauerkraut, drained

1/3 cup minced onion

1/2 cup brown sugar

1 tsp. caraway seed

salt and pepper to taste

2 cups cooked corned beef, diced

8 or 10 rye crackers, crumbled

1/2 pound shredded Swiss cheese

Prepare sauce by melting margarine in saucepan over medium heat. Add flour very quickly while stirring. Add milk slowly, lower heat and stir in mustard. Cook until thickened. Set aside.

COOK POTATOES IN BOILING SALTED WATER ABOUT 10 MINUTES. Drain and set aside.

IN LIGHTLY GREASED 9x13 BAKING DISH, layer half of the sauerkraut, all of the onion and sprinkle over this half of the brown sugar and 1/2 tsp. caraway seed. Next, layer all of the potatoes evenly. Season with salt and pepper as preferred. The third layer is the cooked corned beef. Top with remaining sauerkraut. Distribute the remainder of brown sugar and 1/2 tsp. of caraway seed. Pour sauce over all and top with crumbled rye crackers and all of the shredded cheese. Bake in a 350° oven approximately 25 to 30 minutes. Cool slightly and cut into squares to serve.

Reuben Sandwiches

Yield: 4 Reuben Sandwiches

INGREDIENTS:

8 slices sandwich rye bread

1 stick butter or margarine

12 slices deli corned beef

8 slices Swiss cheese

1 14 oz. can Bavarian kraut, drained

1000 Island salad dressing, to taste

SPREAD BUTTER OR MARGARINE ON ONE SIDE of each slice of bread. Place four slices, buttered side down, in large non-stick skillet over low to medium heat. Place one slice of corned beef on each piece of bread, then about 1/4 of the kraut, 1000 Island dressing and 2 slices of Swiss cheese. Add another slice or two of corned beef and top with another piece of bread (buttered side up).

GENTLY TURN SANDWICH OVER WITH SPATULA to brown the top piece of bread.

Note:
This famous sandwich is said to have originated at the former Blackstone Hotel in Omaha, Nebraska.

Barbequed Pork Back Ribs

Yield: 4 servings

INGREDIENTS:

2 slabs of pork back ribs

salt and pepper to taste

1 18 oz. bottle barbeque sauce, any brand

CUT SLABS OF PORK RIBS IN HALF AND PAR-BOIL for about 25 to 30 minutes. Remove from the water, cut in 2-rib sections and layer in roaster pan. Add salt and pepper.

BASTE WITH BARBEQUE SAUCE at intervals while baking in a 325° oven for about an hour or until ribs are tender.

Crock Pot Sausage and Cabbage

Yield: 6 to 8 servings

INGREDIENTS:

2 pounds smoked Polish sausage

1 medium head cabbage, core removed and coarsely chopped

1 medium onion, chopped

3 to 4 potatoes, peeled and diced

salt and pepper to taste

1/2 tsp. caraway seed

1 15 oz. can chicken broth

1/2 cup water

CUT EACH SAUSAGE INTO 4 PIECES AND TRIM ENDS.

Put chopped cabbage in crock pot. Add onion and diced potatoes.

Add salt, pepper and caraway seed and mix well.

Place cut sausage on top and pour chicken broth and water over all.

Cook on low about 3 hours.

SAUSAGE AND CHEESE CASSEROLE

Yield: about 6 servings

INGREDIENTS:

1-1/2 pounds smoked Polish sausage

5 medium potatoes, peeled and sliced

1 tsp. parsley flakes

salt and pepper to taste

1/2 cup diced onion

2 Tbsp. butter or margarine

2 Tbsp. flour

2 cups milk

2 cups shredded cheddar cheese

CUT POLISH SAUSAGE DIAGONALLY INTO TWO-INCH PIECES. Combine with potatoes, parsley flakes and salt and pepper in greased casserole dish.

TO PREPARE SAUCE: Cook onion in butter or margarine until tender. Stir in flour and add milk all at once, then bring to a boil, stirring to thicken.

POUR SAUCE OVER POTATOES, sausage and seasonings and top with cheddar cheese.

BAKE, COVERED, in a 325° oven for about an hour. Remove lid and return to oven to brown a few minutes more.

Sausage and Pepper Bake

Yield: 4 to 6 servings

INGREDIENTS:

6 links Italian sausage, casings removed

2 medium green peppers, chopped

1/2 cup chopped onion

1 8 oz. can tomato sauce

4 medium potatoes, peeled and diced

salt and pepper to taste

1 or 2 cloves garlic, minced

1/2 tsp. ground oregano

1/2 cup water

BROWN AND CRUMBLE SAUSAGE IN A SKILLET ABOUT 10 OR 15 MINUTES. Remove cooked sausage, combine with remaining ingredients and put in a lightly greased casserole dish.

COVER AND BAKE AT 325° FOR APPROXIMATELY ONE HOUR. Remove casserole lid for the last 15 minutes of baking time.

Sausage and Sweet Potato Casserole

Yield: 6 to 8 servings

INGREDIENTS:

1-3/4 pounds bulk pork sausage

6 medium sweet potatoes

4 to 5 large apples, peeled, cored and sliced

1/2 tsp. ground cinnamon

1 Tbsp. granulated sugar

1/2 cup butter or margarine

1/3 cup brown sugar

FORM PORK SAUSAGE INTO 8 PATTIES. Brown in a skillet on top of the stove and drain on paper towels.

BOIL OR BAKE SWEET POTATOES UNTIL NEARLY TENDER. Then cool just enough to peel and slice.

COVER THE BOTTOM OF A GREASED CASSEROLE DISH with half of the sweet potato slices. Place sausage patties on the sweet potatoes.

MIX APPLE SLICES WITH CINNAMON AND GRANULATED SUGAR and add to the casserole. Now layer on the remaining sweet potato slices adding pats of butter or margarine, and top with brown sugar.

Bake in a 325° oven for 30 to 45 minutes.

Chicken Fried Steak

Yield: 4 to 6 servings

INGREDIENTS:

1-1/2 pounds round steak, tenderized

salt, black pepper and garlic powder to taste

1-1/2 cups flour

1/2 cup milk added to 2 beaten eggs

1 cup seasoned bread crumbs

4 Tbsp. margarine

CUT STEAK INTO SERVING PIECES. Season to taste, dust with 1/2 cup of flour, then dip steak into egg and milk batter. Next, coat with a mixture of remaining flour and seasoned bread crumbs. Allow to dry on floured plate for a few minutes. Brown steak in margarine over medium heat. Lower heat, add a little water and simmer gently for about 30 minutes or until tender.

TURN STEAK ONCE OR TWICE DURING COOKING TIME. Add a small amount of cornstarch dissolved in a little cold water to thicken the pan juices for gravy.

Breaded-Minute Steak

Yield: 4 servings

INGREDIENTS:

4 minute steaks

salt and pepper to taste

1/2 tsp garlic powder

1 cup flour

1/2 cup milk added to 2 beaten eggs

1/2 cup vegetable oil

1 onion, sliced

SEASON MINUTE STEAKS, coat with flour, then dip in egg and milk batter, and again in the flour.

BROWN IN OIL on top of the stove. When brown on both sides, eliminate some of the fat if needed, and add a small amount of water and the onion. Cover skillet, reduce heat and continue cooking on low until tender—about 20 to 30 minutes. Turn steaks at least once during this time.

Minute Steak Skillet Supper

Yield: 4 servings

INGREDIENTS:

4 minute steaks

salt, pepper and garlic powder to taste

1/4 cup flour

3 Tbsp. margarine

1 medium onion, chopped

2 russet potatoes, unpeeled and halved lengthwise

1 large green pepper, seeded and sliced

1 pkg. brown gravy mix, prepared following package instructions

SEASON MINUTE STEAKS—dust with flour and sauté in margarine in a skillet on top of the stove. Add a small amount of water, cover skillet and continue cooking on low approximately 15 minutes. Add onion and potatoes (cut side down) and continue simmering for an additional 15 or more minutes. Add water if needed. Toss in green pepper and pour prepared gravy mix over all. Continue cooking until potatoes are done and steaks are tender.

Salisbury Steak

Yield: 6 servings

INGREDIENTS:

2 pounds ground beef

16 ounces fresh mushrooms, sliced

2 Tbsp. butter or margarine

1 envelope Lipton Onion Soup Mix

1 cup seasoned bread crumbs

2 eggs, beaten

1 Tbsp. Worcestershire sauce

1 Tbsp. garlic powder

1 Tbsp. parsley flakes

1-1/2 Tbsp. coarse ground black pepper

1 cup water

1 pkg. brown gravy mix

SAUTÉ MUSHROOMS in butter or margarine in a skillet on top of the stove.

Remove from skillet. Chop one half of the mushrooms for filling. Remaining sliced mushrooms are added to the gravy.

IN A LARGE BOWL, MIX GROUND BEEF, onion soup mix, bread crumbs, eggs, Worcestershire sauce, garlic powder, parsley flakes, black pepper, 1 cup of water and the chopped mushrooms.

FORM THE MEAT MIXTURE INTO 6 OVAL PATTIES and place in a covered baking pan. Bake about 30 to 40 minutes in a 325° to 350° oven. Take baking pan out of oven and transfer most of the juices to a sauce pan on top of the stove. To this add a little water, the brown gravy mix and a teaspoon or two of cornstarch (mixed first with a little cold water.) Stir and simmer gravy over low to medium heat and when thickened, add the remaining mushrooms. Spoon gravy over the meat patties and return pan to oven for about 10 to 15 minutes.

"Philly" Steak Sandwiches

Yield: 6 sandwiches

INGREDIENTS:

1-1/2 pounds sirloin tip cut in strips

3 Tbsp. margarine

1 green pepper, seeded, and cut julienne style

1 red pepper, seeded, and cut julienne style

1 large onion, sliced

6 hoagie buns

6 slices provolone cheese or mozzarella cheese

SAUTÉ STEAK IN MARGARINE ON TOP OF THE STOVE. When brown, remove from skillet and keep warm. Adding margarine as needed, sauté vegetables until just tender.

SERVE THE STEAK, PEPPERS AND ONION on warm split hoagie buns.

Top with sliced cheese.

Oven Baked Stew

Yield: 6 to 8 servings

INGREDIENTS:

2 pounds beef stew meat

1 pound carrots, peeled and sliced

1 large onion, chopped

2 cups chopped celery

5 potatoes, peeled and cut in chunks

2 14 oz. cans diced or stewed tomatoes

1 pkg. beef stew seasoning mix

1 tsp. garlic powder

1 tsp. coarse ground black pepper

3 Tbsp. Minute Tapioca

1-1/2 cups water

COMBINE ALL INGREDIENTS IN A ROASTER PAN. Mix well and cover pan tightly with extra heavy foil. Place in a 300° oven and bake for about 2 hours. At this time, remove roaster from oven and CAREFULLY REMOVE FOIL. Check tenderness of the meat and stir, adding additional water if needed.

COVER ROASTER PAN and return to oven for another 45 minutes to one hour, or until meat is completely tender. The vegetables will also be done at this time.

Swiss Steak and Dumplings

Yield: about 8 servings

INGREDIENTS:

2 pounds round steak, tenderized

black pepper to taste

1 tsp. garlic powder

1 cup flour

3 Tbsp. margarine

1 cup chopped onion

1 14 oz. can stewed tomatoes

1 8 oz. can tomato sauce

CUT ROUND STEAK INTO SERVING PIECES. Season with garlic powder and pepper and flour lightly. Heat margarine in a skillet and brown meat. Transfer the steak to kettle or Dutch oven and add chopped onion. Add the stewed tomatoes and tomato sauce.

COVER AND SIMMER, TURNING OCCASIONALLY, on low heat for about an hour or until steak is tender. Add a small amount of water (or more tomato sauce if needed) to skillet during cooking time.

PREPARE EASY POTATO DUMPLINGS found on page 144 of the "Potatoes" section.

TRANSFER STEAK AND DUMPLINGS TO SERVING PLATTER AND POUR PAN JUICES OVER ALL.

Veal Parmesan

Yield: 4 servings

INGREDIENTS:

4 veal cutlets

salt and pepper to taste

Italian seasoning to taste

1/2 cup flour

1/2 cup milk added to 2 beaten eggs

1 cup seasoned bread crumbs

1/4 cup additional flour

1/2 cup vegetable oil

2-1/2 cups spaghetti sauce or marinara sauce

2 cups mozzarella cheese, shredded

SEASON AND FLOUR CUTLETS, DIP IN MILK AND EGG BATTER and then coat with a mixture of additional flour and seasoned bread crumbs.

SAUTÉ IN OIL until brown on both sides. Add a little water to the skillet and cook on low for about 15 minutes—turn cutlets once or twice.

INTO A BAKING PAN OR ROASTER PAN, POUR ABOUT 2 cups spaghetti sauce or marinara sauce. Add the cutlets and top with more sauce.

BAKE, COVERED, IN A 300° OVEN for about 20 minutes. Remove from oven, add a generous amount of shredded mozzarella, and return (uncovered) to oven until cheese melts. Serve with cooked pasta and heated spaghetti or marinara sauce.

VARIATION:
BREADED VEAL CUTLETS (SCHNITZEL)

SEASON WITH SALT AND PEPPER AND FLOUR TENDERIZED CUTLETS, DIP IN EGG AND MILK BATTER and coat with flour and bread crumbs as above. Sauté the cutlets in vegetable oil. Serve with cooked noodles and Sour Cream-Dill Gravy on page 193 in the "Sauce and Gravy" Section.

Veal Shanks-Osso Buco

Yield: 4 servings

INGREDIENTS:

2 pounds veal shanks cut 1-1/2 inch thick

black pepper to taste

garlic powder to taste

1/2 cup flour

3 Tbsp. butter or margarine

1 cup sliced carrots

1 cup chopped onion

1/2 cup peeled and sliced parsnips

2 cups chicken stock

1/2 cup tomato paste

1/2 cup white wine

1 Tbsp. fresh rosemary

1/8 tsp. ground cloves

1/8 tsp. tarragon leaves

SEASON VEAL SHANKS WITH BLACK PEPPER AND GARLIC POWDER and dust with flour. Brown the shanks in margarine in a large skillet on top of the stove. Remove shanks and place in a roaster pan.

IN SAME SKILLET ADD MORE MARGARINE IF NEEDED AND SAUTÉ the vegetables, about 10 minutes. Add chicken stock, tomato paste, wine, rosemary, ground cloves and tarragon leaves. Simmer over medium heat until reduced by half. Pour the reduced sauce over the veal shanks in roaster pan. Cover and bake in a 300° oven for approximately one hour or until veal is tender. To serve, place the veal on a platter, and spoon sauce over all.

Note:
"Osso Buco" translates as 'hole of bone,' and refers to the rich flavor derived from the veal bone marrow.

Spicy Wieners and Beans

Yield: 6 to 8 servings

INGREDIENTS:

6 or 8 natural casing wieners, sliced diagonally

2 15 oz. cans pork and beans

1 15 oz. can chili with beans

2 Tbsp. brown sugar

2 16 oz. cans large lima beans, drained

1/3 cup catsup

1/2 cup chopped onion

salt and pepper to taste

1/8 tsp. crushed red pepper

4 slices hickory smoked bacon, cut in thirds

IN A 3-QUART CASSEROLE, COMBINE: wieners, pork and beans, chili, brown sugar, lima beans, catsup, onion, salt, pepper and crushed red pepper.

TOP WITH BACON PIECES and bake in a 325° oven for 45 to 50 minutes. Stir once or twice while baking and again before serving.

The Outdoor Grill

Grilling or cooking outdoors has taken place for centuries; however, this method of suburban meal preparation really gained popularity in the 1950's, with the introduction of the first "Weber" grill.

My first recollection of grilling outdoors was probably in about 1953, when a favorite aunt cooked hamburger patties on a simple, inexpensive charcoal grill. The burgers were delicious, and quite a novelty at the time.

Beef and Cheese Burgers and Taco Burgers will add a little pizzazz to the everyday hamburger cookout. Polynesian Grilled Chicken and Polish Sausage Luau were inspired by our 25th anniversary trip several years ago to the Hawaiian Islands.

Beer Bratwurst

Yield: 6 servings

INGREDIENTS:

6 links fresh bratwurst

MARINADE:

1 12 oz. can beer

1 tsp. chili powder

1 Tbsp. prepared mustard

2 Tbsp. brown sugar

1 tsp. instant minced garlic

small amount of Tabasco sauce

COMBINE BEER AND ALL OTHER INGREDIENTS in glass bowl or dish. Place bratwurst in the marinade and turn once or twice while refrigerating several hours or overnight. Remove sausage and discard marinade.

GRILL SAUSAGE OVER MEDIUM-HOT COALS about 10 to 15 minutes per side or until done. Turn occasionally while cooking.

SERVE BRATWURST in small sausage buns or hoagie buns. Top with sauerkraut or stone ground mustard.

Beef and Cheese Burgers

Yield: 6 servings

INGREDIENTS:

2-1/2 pounds ground beef

1 cup Swiss cheese, shredded

garlic powder to taste

grated Parmesan cheese to taste

12 slices hickory smoked bacon

coarse ground black pepper to taste

Lawry's seasoned salt to taste

DIVIDE GROUND BEEF AND FORM INTO TWELVE 3-INCH PATTIES.

PLACE ABOUT ONE TABLESPOON OF THE SHREDDED CHEESE ON 6 of the patties and shake on garlic powder and Parmesan cheese. Cover with the remaining patties and seal edges. Wrap each patty with 2 slices of bacon and secure with wooden picks.

SEASON PATTIES ON BOTH SIDES with pepper and seasoned salt and cook on charcoal or gas grill approximately 10 to 12 minutes per side, or until juices run clear.

TACO BURGERS

Yield: 8 burgers

INGREDIENTS:

1-1/2 pounds ground beef

1/2 cup seasoned bread crumbs

1 15 oz. can refried beans

1/2 cup diced onion

1 4 oz. can chopped green chiles

1 tsp. ground cumin

3/4 Tbsp. coarse ground black pepper

1 cup cheddar cheese, shredded

8 flour tortilla shells

TOPPINGS:

chopped lettuce

chopped tomatoes

chopped onion

sour cream

salsa or picante sauce

shredded cheddar cheese

COMBINE GROUND BEEF, BREAD CRUMBS, REFRIED BEANS, ONION, CHILES, CUMIN AND COARSE GROUND BLACK PEPPER. Mix well. Form eight thin patties, about 6 inches in size, and distribute a small amount of cheese on each. Fold patties in half to form a semi-circle.

COOK ON GRILL RACK OVER MEDIUM COALS—about 12 to 15 minutes. Aluminum foil or a grill basket may be used. Heat tortilla shells (on foil) on the grill—or use packaged pre-formed corn tortillas. Fill shells with taco burgers.

Top with lettuce, tomato, onion, sour cream, salsa or picante sauce and more shredded cheese if desired.

Glazed Chicken On The Grill

Yield: 6 to 8 servings

INGREDIENTS:

2 whole, fresh broiler-fryer chickens

GLAZE:

1/2 cup margarine

3/4 cup light corn syrup

1/2 cup lemon juice

1 tsp. sage

1 tsp. paprika

2 tsp. ground oregano

2 tsp. Italian seasoning

1/2 tsp. garlic salt

1 tsp. dill weed

1 tsp. coarse ground black pepper

1/2 cup grated Parmesan cheese

INSERT SPIT ROD THROUGH CHICKENS, SECURING WITH HOLDING FORKS ON EACH END. For proper balance, tie twine around each bird securing wings to body—legs should also be fastened to spit rod. Attach the spit rod to the grill (grill racks removed) over medium coals. Brush chickens lightly with melted butter or margarine. Turn on motor and begin roasting chickens on the spit rod. Adjust temperature of coals so that a medium to low heat can be maintained. Lower hood on grill and continue cooking on spit rod for approximately one hour.

FOR GLAZE, MELT MARGARINE IN SAUCEPAN and add all remaining ingredients. Simmer a few minutes over low heat. After chickens have cooked for about an hour, begin basting with glaze every 15 minutes. Continue cooking for about 30 to 45 minutes, or until a meat thermometer inserted into the breast portion of chicken registers 165° to 170°.

If flare-up occurs, use a water spray bottle to control flames, and/or place a drip pan below the rotisserie rod.

Oriental Chicken Bundles

Yield: 6 servings

INGREDIENTS:

3 whole bone-in chicken breasts, cut in halves

black pepper and garlic salt to taste

3 cups quick cooking rice, raw

1 pound fresh mushrooms, sliced

2 cups snow peas

soy sauce as desired

ground ginger

1 bunch green onions, sliced

9 Tbsp. margarine

4 cups hot water

extra heavy aluminum foil

SEASON CHICKEN BREASTS WITH BLACK PEPPER AND GARLIC SALT AND BROWN ON LIGHTLY GREASED GRILL over medium to hot coals. This requires 15 to 20 minutes of cooking time. Set aside. Tear off six 18-inch square sheets of EXTRA HEAVY FOIL. In the center of each sheet, spread 1 Tbsp. margarine. Next, add 1/2 cup rice, mushrooms and snow peas. Shake on some soy sauce. Place one chicken breast on each, and sprinkle on a pinch of ground ginger. Now add more soy sauce, green onion and 1/2 Tbsp. margarine. Bring corners of the foil together to partially form each pouch, then slowly pour 2/3 cup hot water into the pouch. Finally, close tightly and seal tops of each. Place chicken bundles on grill over low to medium coals and cook about 25 to 30 minutes, or until rice is done and chicken registers 165° to 170° on meat thermometer.

Note:
This recipe works best if prepared on a covered grill. Whether covered or open grill, individual pouches should be checked during cooking time to make certain there is enough liquid in each to cook the rice. Add more water (or soy sauce) if needed.

Serve with: Bacon-Onion Rolls

Bacon-Onion Rolls

Yield: about 12 Bacon-Onion Rolls

INGREDIENTS:

2 pkgs. crescent dinner rolls

6 strips hickory smoked bacon

1 cup chopped green onion

butter or margarine, melted

FRY BACON UNTIL CRISP. Drain on paper towels and crumble. Set aside.

ON FLOURED SURFACE, unroll both packages of crescent rolls and pinch together all sections until no perforations remain. Brush with melted butter or margarine. Distribute evenly the green onion and crumbled bacon.

ROLL UP jelly-roll fashion and cut at 1-1/2 to 2 inch intervals. Flatten each 'slice' with a rolling pin and place on lightly greased cookie sheet.

BRUSH TOPS OF ROLLS WITH ADDITIONAL BUTTER OR MARGARINE and bake in a 325° to 350° oven approximately 10 to 12 minutes.

Polynesian Grilled Chicken

Yield: 6 to 8 servings

INGREDIENTS:

2 chickens, quartered

MARINADE:

1 15 oz. can crushed pineapple, reserve juice

1 cup soy sauce

1/2 cup vegetable oil

1 cup white wine

2 tsp. curry powder

1 cup brown sugar

1/2 tsp. dry mustard

1 tsp. garlic, minced

1 tsp. coarse ground black pepper

2 tsp. sesame seed

IN A BLENDER, MIX pineapple, soy sauce, vegetable oil, white wine, curry powder, brown sugar and dry mustard. Pour mixture into a large glass dish and add minced garlic, black pepper and sesame seed.

PLACE QUARTERED CHICKENS in the dish or bowl and marinate in the refrigerator at least 4 to 6 hours or overnight. Turn chicken once or twice during this time. Discard left-over marinade.

COOK CHICKEN ON LIGHTLY GREASED GRILL OVER LOW TO MEDIUM COALS. Continue grilling approximately 45 minutes to one hour, or until meat thermometer registers 165° to 170° when inserted into a breast portion of the chicken. Baste during cooking time with a little melted margarine and reserved pineapple juice and additional soy sauce.

FOR A SPECIAL TOUCH:

Place hot, cooked chicken on a serving platter and top with one cup of shredded or flaked coconut. Arrange fresh sliced pineapple around edge of platter.

BREADS FOR THE GRILL, TWO WAYS

Yield: about 8 servings

INGREDIENTS:

GARLIC GRILL LOAF:

1 loaf cook-out bread or uncut French bread

butter or margarine, melted

Italian seasoning

garlic powder

grated Parmesan cheese

GRILLED RYE LOAF:

1 loaf uncut light rye bread

1 stick butter or margarine

3/4 cup shredded Swiss cheese

fresh snipped parsley or parsley flakes

MAKE DIAGONAL CUTS IN THE BREAD, being careful not to cut all the way through the loaf. With a pastry brush, coat bread between cuts with melted butter or margarine and shake on garlic powder and Italian seasoning.

BRUSH MORE BUTTER OR MARGARINE ON TOP and shake on grated Parmesan cheese. Wrap the loaf loosely in foil and place on grill rack (not directly over hot coals) to heat.

SLICE BREAD DIAGONALLY, WITHOUT CUTTING THROUGH THE LOAF. Spread butter or margarine, Swiss cheese and parsley between the slices. Wrap the loaf in foil and place on the coolest part of the grill for 10 to 12 minutes. Cut through slices or pull apart to serve.

Whole Ham On the Grill

Yield: 12 servings

INGREDIENTS:

1 hickory smoked bone-in ham

2 tsp. ground cloves

SCORE TOP OF HAM IN A CRISSCROSS PATTERN and rub in the ground cloves. Place ham on lightly greased grill rack which has been raised to highest position over low-glowing coals. A heavy metal drip pan directly below the ham—and with coals banked around it—is a must!

LOWER LID OF COVERED GRILL and roast ham for approximately 1 hour. Lift grill hood and begin basting with a glaze or sauce. See Orange-Pineapple Glaze recipe on page 75.

Total cooking time should be approximately 1-1/2 hours, or until meat thermometer inserted into the ham registers about 140°.

Polish Sausage Luau

Yield: 4 to 6 servings

INGREDIENTS:

3 links smoked Polish sausage (each cut into 6 or 8 diagonal pieces)

2 large green peppers, cut into chunks

2 20 oz. cans pineapple chunks, drained

3 or 4 sweet potatoes, partially cooked, peeled and cut into chunks

6 wooden skewers

GLAZE:

2/3 cup pineapple preserves

1/2 cup brown sugar

1/4 cup lemon juice

1 Tbsp. Dijon mustard

2 Tbsp. Worcestershire sauce

SIMMER PINEAPPLE PRESERVES, BROWN SUGAR, lemon juice, mustard and Worcestershire sauce in a saucepan about 5 minutes, and stir once or twice.

THREAD CUT SAUSAGE ON SKEWERS alternating with sweet potatoes, pineapple and green pepper.

BASTE WITH GLAZE while cooking over medium coals, turning frequently, for 10 to 12 minutes.

POTATOES IN A FOIL POUCH

Yield: 6 to 8 servings

INGREDIENTS:

6 russet potatoes

1 large onion, sliced

salt and pepper to taste

1 to 1-1/2 cups boiling water

6 Tbsp. butter or margarine

2 tsp. parsley flakes

extra heavy foil

WASH AND SLICE RUSSET POTATOES. Potatoes may be peeled or skins left on.

Put potato slices and onion on a greased extra heavy sheet (doubled) of aluminum foil—folding up edges on all sides. Add salt and pepper and pour boiling water over potatoes and onions in the pouch. Top with butter or margarine and parsley flakes. To seal pouch, bring up and pinch together opposite corners of the foil.

COOK OVER LOW GLOWING COALS ON GRILL RACK— about 20 minutes total cooking time.

Teriyaki Barbeque Ribs

Yield: 6 to 8 servings

INGREDIENTS:

4 slabs pork back ribs, cut into halves

BARBEQUE SAUCE:

1 15 oz. can tomato sauce

1 6 oz. can tomato paste

3/4 cup brown sugar

1/2 cup soy sauce

1/4 cup cider vinegar

1/2 cup minced onion

1 tsp. chili powder

1 tsp. each black pepper, garlic salt and crushed sweet basil leaves (or fresh basil)

1 Tbsp. ground ginger

1 Tbsp. whole mustard seed

1 Tbsp. Kitchen Bouquet

PRE-COOK RIBS IN BOILING WATER ABOUT 15 TO 20 MINUTES.

PLACE PARTIALLY COOKED RIBS ON LIGHTLY GREASED GRILL over low-glowing coals. A drip pan may be placed in the center of coals.

IN A BOWL, combine all barbeque sauce ingredients and stir well.

BASTE RIBS LIGHTLY WITH SAUCE. Turn occasionally and continue basting until ribs are tender—approximately 1-1/2 to 2 hours. Ribs may be placed on sheets of heavy aluminum foil for a portion of the cooking time.

Camper's Stew

Yield: 6 to 8 servings

INGREDIENTS:

2 pounds smoked Polish sausage

5 medium russet potatoes, peeled and quartered

5 or 6 carrots, cleaned and thickly sliced

3 green peppers, seeded and cut in chunks

2 onions, cut in chunks

1 16 oz. bag frozen corn

2 Tbsp. margarine

salt and pepper to taste

garlic powder to taste

1 cup prepared BBQ sauce

3/4 cup water

extra heavy foil

CUT EACH SAUSAGE INTO 5 OR 6 PIECES AS PREFERRED. Tear off two large sheets of EXTRA HEAVY FOIL and place one on the other (vertical and horizontal). Spray with vegetable coating spray. Begin with potatoes, and add carrots, onion, green pepper, frozen corn and seasonings. Add the Polish sausage and BBQ sauce. Top with margarine and pour in the water. Fold over one sheet of the foil to make a center seam and close. Ends of the second sheet should next be secured and brought over the top of package.

PLACE FOIL PACKAGE ON GRILL RACK OVER LOW TO MEDIUM COALS. Grill about 45 minutes, or until potatoes and carrots are tender. The foil package should be opened once or twice during cooking time to stir ingredients. Add a small amount of additional water or BBQ sauce if needed.

Fish and Seafood

    Fish was a familiar Friday entrée for years in our home—Tuna and Noodle Casserole and Salmon Patties were popular meal choices, and now and then, there was the special treat of a fast-food fish sandwich or a shrimp take-out meal.

    Fresh caught fish was always a favorite, whether it was catfish, sunfish, or the walleye, which we enjoyed during our summer vacation trips to Minnesota.

    My recipe for Shrimp Creole came about after a visit to New Orleans. This tasty dish is seasoned with Filé powder, which is made from dried and ground Sassafras leaves. The Filé seasoning adds both flavor and thickening, and is frequently used in Cajun Cuisine.

Crab and Shrimp Cakes

Yield: 6 servings

INGREDIENTS:

2 6 oz. cans canned crab meat, drained

1 4 oz. can tiny whole shrimp, drained and rinsed

2 eggs, beaten

1/4 cup chopped onion

1/4 cup chopped green pepper

1/4 cup chopped red pepper

1/4 cup chopped celery

1-1/2 cups seasoned bread crumbs

1/2 cup mayonnaise

1 tsp. lemon pepper seasoning

1/2 tsp. garlic powder

1/4 tsp. dill weed

1/4 tsp. paprika

1/4 tsp. crushed red pepper

small amount of flour

4 Tbsp. vegetable oil

COMBINE CRAB MEAT, SHRIMP, eggs, onion, green and red pepper, celery and bread crumbs. To this mixture, add mayonnaise and all seasonings and mix well.

FORM MIXTURE INTO SIX PATTIES and coat each with a little flour. Sauté in vegetable oil until brown on both sides.

Breaded Fish Filets

Yield: 3 or 4 servings

INGREDIENTS:

1 pound or more fish filets, fresh or frozen

1/2 cup flour

2 eggs

1/2 cup milk

1/4 cup bread crumbs

1/4 cup flour

2 Tbsp. instant potato flakes

salt and pepper to taste

1/4 tsp. dill weed

1/4 cup vegetable oil

THAW FISH IF FROZEN—blot on paper towels. Salt and pepper to taste.

PLACE 1/2 cup flour in a shallow bowl. MIX the eggs and milk in a second shallow bowl. COMBINE bread crumbs, flour, potato flakes and dill in a third shallow bowl.

Dip fish filets in flour, next in the mixture of egg and milk, and then press into the bread crumb mixture to coat.

FRY IN VEGETABLE OIL JUST UNTIL BROWNED ON BOTH SIDES.

Microwave Fish Filets

Yield: 3 to 4 servings

INGREDIENTS:

1 pound fish filets, torsk or other white fish

2 Tbsp. melted butter or margarine

1/2 cup lemon juice

lemon pepper seasoning, to taste

1/2 tsp. dill weed

PLACE THE FISH FILETS IN A GLASS CASSEROLE DISH. Add melted butter or margarine to the lemon juice and pour over fish. Shake on the lemon-pepper seasoning and dill weed.

COVER WITH PLASTIC WRAP AND COOK IN MICROWAVE OVEN ON HIGH about 3 to 4 minutes. Turn dish a quarter turn, and microwave for about 5 minutes more.

Salmon Patties

Yield: about 6 Salmon Patties

INGREDIENTS:

1 14 oz. can pink salmon, drained (bone removed)

3/4 cup seasoned bread crumbs

2 eggs, beaten

1/2 cup finely chopped celery

1/3 cup finely chopped onion

1/4 tsp. black pepper

1/2 tsp. dill weed

1/2 tsp. Paul Prudhomme's Seafood Magic

1/4 cup vegetable oil

COMBINE SALMON, bread crumbs, eggs, celery, onion and seasonings in a medium sized bowl. Mix well and form into patties. Mixture should have the consistency of a hamburger patty. (Add a very small amount of water or milk if too dry.)

HEAT VEGETABLE OIL IN A SKILLET—Cook patties, turning until brown on both sides.

Note:
Paul Prudhomme's Seafood Magic is wonderful for all fish or seafood entrées.

Shrimp Creole

Yield: 4 to 6 servings

INGREDIENTS:

2 pounds shrimp, peeled, deveined, and tails removed

3 Tbsp. margarine

1 cup chopped celery

1 medium chopped onion

1 green pepper, seeded and chopped

2 cloves garlic, finely minced

2 14 oz. cans stewed tomatoes

1 Tbsp. Paul Prudhomme's Seafood Magic

1/4 tsp. crushed red pepper

1/4 tsp. thyme leaves

1 tsp. ground oregano

2 tsp. granulated sugar

1/2 tsp. coarse ground black pepper

1/4 tsp. chili powder

2 Tbsp. Filé powder

6 cups cooked rice

MELT THE MARGARINE IN A DEEP SKILLET and sauté the celery, onion, green pepper and garlic until just tender.

ADD THE TOMATOES, the Seafood Magic, crushed red pepper, thyme leaves, oregano, sugar, black pepper and chili powder. Simmer on low about 30 minutes, stirring occasionally.

NEXT, ADD the Filé powder and the raw shrimp. SIMMER GENTLY FOR ANOTHER 10 TO 15 MINUTES—The shrimp requires only a few minutes of cooking time. Serve Shrimp Creole over hot, cooked rice.

Note:
If pre-cooked shrimp is used, add the shrimp only to heat through just before serving.

Tuna and Noodle Casserole

Yield: 6 to 8 servings

INGREDIENTS:

2 6 oz. cans water-pack tuna, drained

1 10 oz. can cream of mushroom soup

1 10 oz. can cream of celery soup

2 eggs, beaten

3/4 cup milk

1/4 cup finely chopped onion

1/4 cup chopped green pepper

1/4 cup chopped red pepper

salt and pepper to taste

1 tsp. dried parsley flakes

1 12 oz. pkg. frozen noodles, cooked and drained

1 to 1-1/2 cups potato chips, crushed

COMBINE the cans of soup with the eggs and stir in the milk. Add onion, green and red pepper, tuna, salt, pepper, parsley flakes, and noodles. Pour into greased 9x13 glass casserole dish and top with crushed potato chips. Bake at 325° to 350° approximately 45 minutes, or until hot and golden brown. Take casserole out of the oven and allow to cool slightly before serving.

Specialty Meat Entrées

The 'Butcher' and I joined the Nebraska Association of Meat Processors in 1972, and we attended numerous State Conferences over the years. This Nebraska trade organization was founded in 1939, and has afforded its members a vast array of knowledge regarding meat and food preparation, as well as current food safety guidelines.

As an Association member, I had the opportunity to create a few gourmet meat entries in the various classes of Specialty Meat Competition. My husband and children were involved in the testing and tasting process, during which I modified recipes, added ingredients, and adjusted methods of preparation.

My entries were all winners of top awards in Specialty Meat contests held throughout the State of Nebraska. And, although many of them are more time consuming to prepare, they are unique and will be well worth the effort for a special occasion dinner.

Apricot Pork Pecan

Yield: 8 servings

INGREDIENTS:

2 2-pound sections boneless pork loin, scored and pounded

1/2 tsp. coarse ground black pepper

1/2 tsp. poultry seasoning

1 pound bulk pork sausage

1 cup seasoned bread crumbs

1 egg white, beaten

1/2 cup dried apricots, chopped

1 cup pecans, chopped

3/4 cup apricot preserves

1 cup dried apricots, (softened)*

1/2 cup whole pecans

butcher twine

SEASON PORK LOIN SECTIONS with black pepper and poultry seasoning.

BROWN PORK SAUSAGE – Crumble as it cooks, then drain fat and cool slightly. Mix sausage with bread crumbs, egg white, chopped apricots and the chopped pecans.

PLACE ONE OF THE PORK LOIN SECTIONS, fat side down on work area. Spread on the sausage mixture. Add the second piece of pork (fat side up) and tie roast at intervals with butcher twine.

PLACE PORK IN ROASTER PAN and bake in a 325° oven for approximately 1 hour. AT THIS TIME, BEGIN BASTING WITH APRICOT PRESERVES until the preserves form a glaze. Remove roast from oven and arrange whole apricots and whole pecans on top of the pork, pressing them into the glaze. Return to oven and baste with preserves while continuing to cook until internal temperature of the roast registers 155° to 160° on meat thermometer.

*To soften dried apricots, place them in a single layer on a microwave safe dish. Sprinkle with a little water, cover and microwave for 1 to 2 minutes.

Mexican Beef Roll

Yield: 8 to 10 servings

INGREDIENTS:

1 4-pound sirloin tip, butterflied and flattened

2 pounds lean ground beef

1 cup bread crumbs

1/2 tsp. coarse ground black pepper

1 tsp. chili powder

2 tsp. ground cumin

2 eggs, beaten

water to moisten

2 cups corn meal

1 4 oz. can chopped green chiles

1 4 oz. can chopped black olives

5 to 6 corn tortillas

1 Tbsp. crushed red pepper

butcher twine

COOK GROUND BEEF, CRUMBLE AND DRAIN EXCESS FAT. Mix in bread crumbs, black pepper, chili powder and 1 tsp. ground cumin. Add beaten eggs and a little water if needed. Mixture should have a firm consistency.

COOK CORNMEAL according to package directions. Cool slightly and add green chiles and chopped black olives.

SPREAD GROUND BEEF MIXTURE ON THE FLATTENED SIRLOIN. Next, layer on the corn tortillas, and spread the cornmeal mixture over tortilla shells. Beginning with the narrow end, roll roast (jelly-roll fashion) and tie with butcher twine – trim ends. On waxed paper, distribute crushed red pepper and 1 tsp. ground cumin. Roll the roast in these seasonings. Place beef roll (seam side down) in a roaster pan, which has been sprayed with vegetable coating spray. Bake in a 325° oven until internal temperature registers 145° on meat thermometer.

ORIENTAL PORK LOIN

Yield: 8 servings

INGREDIENTS:

1 4-pound boneless pork loin butterflied and flattened

black pepper to taste

1-1/2 tsp. Chinese 5-spice powder

1-1/2 pounds bulk pork sausage

1-1/2 cups seasoned bread crumbs

1 egg white, beaten

3 cups cooked white rice

1 clove garlic, minced

1/2 cup chopped green onion

2 egg whites, beaten

8 oz. fresh mushrooms, sliced and sautéed in 2 Tbsp. butter or margarine

1 4 oz. jar chopped pimento, drained

1 cup unsalted cashews, chopped

dried parsley flakes

1 or 2 Tbsp. fresh ginger, chopped

butcher twine

SEASON THE PORK LOIN WITH 1 tsp. Chinese 5-spice powder and black pepper to taste.

BROWN PORK SAUSAGE, then drain excess fat and cool. Mix with bread crumbs and egg white. Spread sausage mixture on the pork loin, which should be fat side down on work area.

TO THE COOKED RICE ADD 1/2 tsp. Chinese 5-spice powder, minced garlic, green onion, egg whites, mushrooms and pimento. Layer this over the pork sausage mixture. Roll, beginning with narrow end and continue until the opposite end is reached. Remove excess filling which has pushed out of the roast and tie at intervals with butcher twine.

ON WAXED PAPER, DISTRIBUTE the cashews and parsley. Coat surface of roast by rolling in these ingredients, pressing in as much as possible. Place pork in roaster pan (fat side up) and top with chopped fresh ginger. Bake in a 325° oven, adding water to the pan if needed and basting occasionally, until an internal temperature of 155° to 160° is reached.

Pork Polynesian

Yield: 8 to 10 servings

INGREDIENTS:

1 4-pound boneless pork loin, butterflied and flattened

1-1/2 pounds bulk pork sausage

2 cups seasoned bread crumbs

2 eggs, beaten

salt and pepper to taste

1/4 tsp. ground ginger

3 fresh sweet potatoes

1 15 oz. can crushed pineapple, drained

1 medium green pepper, chopped fine

butcher twine

GLAZE:

1/2 cup orange marmalade

1/2 cup pineapple preserves

BROWN PORK SAUSAGE – drain excess fat. Mix in the bread crumbs, eggs and seasonings.

COOK SWEET POTATOES in boiling water just until partially cooked. Cool, peel and dice. Add these to the sausage mixture together with the pineapple and green pepper.

IN A SMALL BOWL, combine marmalade and pineapple preserves.

SPREAD MIXTURE ON THE PORK (fat side down.) Roll, beginning with narrow end and continue until the opposite end is reached. Remove excess filling which has pushed out of the roll. Tie roast at intervals with butcher twine and place fat side up in a roaster pan. Bake uncovered at 325° for approximately 20 to 30 minutes. Cover pan and continue baking for 1 hour, basting occasionally with pan juices. Begin to brush on the orange pineapple glaze at intervals and bake for 30 more minutes or until a meat thermometer inserted into the pork reads 155° to 160°.

Pork and Dumpling Roast

Yield: 8 to 10 servings

INGREDIENTS:

1 5-pound boneless pork loin, butterflied and flattened

1-1/2 pounds bulk pork sausage

1-1/2 cups bread crumbs

1 egg, beaten
bamboo skewers and butcher twine

SWEET SOUR CABBAGE:

1 medium cabbage, cored and chopped

1 small onion, chopped

salt and pepper to taste

2 tsp. granulated sugar

2 Tbsp. butter

1 Tbsp. flour

2 Tbsp. white vinegar
1 tsp. caraway seed

BREAD DUMPLINGS:

3 slices dried white bread, crumbled (no crusts)

3 cups flour

3-1/4 tsp. baking powder

1 tsp. salt

2 eggs, well beaten

3/4 cup milk

BROWN SAUSAGE, then drain excess fat and cool. Mix with bread crumbs and beaten egg. Sausage may be cooked ahead and refrigerated. Bring to room temperature before assembling roast.

COOK CABBAGE AND ONION in boiling water for about 5 minutes or until barely tender and drain. Add salt and pepper, sugar, butter, flour, vinegar and caraway seed. Note: Sweet-sour cabbage may also be prepared ahead and refrigerated; however, you will need to bring to room temperature before assembling roast.

COMBINE DRY INGREDIENTS with egg and milk and form into a soft dough on a floured board. With floured hands, shape the dough into 3 or 4 rolls that are about 8 inches long and 2 inches wide. Allow to rise under a covered dish for at least 30 minutes. Cook the dumplings in boiling water for 8 to 10 minutes. Remove from boiling water onto a rack or platter and allow to cool. DUMPLINGS SHOULD NOT BE PREPARED AHEAD. Two of the rolls are used for the roast…serve the remainder with gravy.

Pork and Dumpling Roast (continued)

TO ASSEMBLE ROAST:

Place the pork loin on cutting board or work area, fat side down, and spread the sausage mixture to within a half-inch of the edges of the pork loin. Next, place dumpling rolls down the center (as a rule, two rolls placed end to end are needed.) Spoon generous amounts of the sweet-sour cabbage on either side of the dumplings.

BEGINNING WITH THE SHORT SIDE OF THE LOIN, roll the roast, jelly-roll fashion, until the end is reached. Skewer at intervals to fasten the roll, and tie the roast with butcher twine. Excess filling which has pushed out of the roast should be removed. Smooth ends and remove skewers. Roll the roast in caraway seed, pressing in as much as possible.

SPRAY THE BOTTOM OF A LARGE ROASTER PAN with a vegetable coating spray and place the roast in the pan (fat side up.) Add a little water to the pan and bake uncovered at 350° for approximately 20 to 30 minutes. Cover pan with lid and continue baking for about 1-1/2 hours at 300°. Baste occasionally with pan juices, adding water to the pan now and then if needed. Roast is done when meat thermometer registers 155° to 160°. Gravy may be made with pan drippings following All Purpose Gravy instructions on page 190-191 in "Sauce and Gravy" Section.

Steak and Shrimp Pinwheels

INGREDIENTS:

skirt steak, cut into strips

jumbo shrimp or prawns, peeled and tails removed

butter or margarine, melted

bottled Italian salad dressing

WRAP STEAK STRIPS AROUND EACH SHRIMP OR PRAWN beginning at center, and fasten with wooden picks in a crisscross fashion. Pinwheels must be securely fastened and should lie flat.

BRUSH WITH A MIXTURE OF MELTED BUTTER OR MARGARINE AND ITALIAN DRESSING, and grill over medium coals on greased grill rack— approximately 4 to 6 minutes per side.

STEAK AND SHRIMP PINWHEELS MAY ALSO BE PREPARED BY BROILING IN THE OVEN.

Note:
Shake on coarse ground black pepper or lemon pepper seasoning if desired.

POTATOES

There are numerous varieties of potatoes; and, whether baked, scalloped, mashed or fried, potatoes are the ultimate comfort food.

I prepared potatoes often for my husband and children and some of their favorites, in addition to mashed potatoes, were Pizza Potatoes, Baked Italian Fries and a Twice-Baked Potato Casserole.

The recipe for Potato Pancakes is similar to the one that my father prepared as part of his Friday meatless dinner menu.

BAKED ITALIAN FRIES

Yield: 6 to 8 servings

INGREDIENTS:

6 russet potatoes, washed but not peeled

2 sticks butter

1 to 2 pkgs. Good Seasons Italian salad dressing mix

COOK WHOLE POTATOES IN BOILING WATER UNTIL NEARLY TENDER. Cool and cut the potatoes into halves lengthwise. Then cut each half into thirds lengthwise again, leaving the skins on.

PLACE POTATO WEDGES ON A BAKING SHEET which has been coated with vegetable spray. Melt the butter and brush on the potatoes. Sprinkle on 1 or 2 packages dry Italian dressing mix. Bake in a 325° oven about 20 minutes or until hot and slightly browned.

'Better' Au Gratin Potatoes

Yield: 4 to 6 servings

INGREDIENTS:

1 box Au Gratin Potatoes

1/2 cup finely minced onion

1/2 tsp. coarse ground black pepper

1 Tbsp. dried parsley flakes

1/4 tsp. garlic powder

MIX AU GRATIN POTATOES ACCORDING TO PACKAGE DIRECTIONS—Add onion, black pepper, parsley flakes and garlic powder. Bake as directed on package.

Easy Potato Dumplings

Yield: 12 to 15 dumplings

INGREDIENTS:

2-1/2 cups Bisquick baking mix

1 cup instant potato flakes

salt and pepper to taste

parsley flakes (optional)

1 cup milk

3/4 cup flour

1 or 2 chicken bouillon cubes

MIX TOGETHER THE BISQUICK, POTATO FLAKES, AND SEASONINGS. Slowly add milk and mix thoroughly.

FORM MIXTURE INTO ROUND DUMPLINGS WITH A TABLESPOON and roll each in a little flour. Drop them into a kettle of boiling water to which a chicken bouillon cube or two has been added. Lower heat and cook the dumplings about 10 minutes. Cover and cook an additional 5 to 10 minutes.

WITH A SLOTTED SPOON, TRANSFER DUMPLINGS to a warm platter and serve with gravy; or cool slightly and brown in a skillet with chopped onion and margarine.

Hash Brown Potato Bake

Yield: 8 or more servings

INGREDIENTS:

- 1 32 oz. pkg. frozen hash brown potatoes
- 1 10 oz. can cream of mushroom soup
- 1 10 oz. can cream of celery soup
- 1 egg, beaten
- 1/2 cup milk
- 1/2 cup minced onion
- salt and pepper to taste
- 2 tsp. parsley flakes
- 6 Tbsp. butter or margarine
- 1 pint sour cream
- 2 cups cheddar cheese, shredded

PARTIALLY THAW POTATOES AND COMBINE IN A BOWL with the creamed soups, egg mixed with milk, onion, salt and pepper and parsley flakes.

TURN THE MIXTURE INTO A GREASED 9x13 CASSEROLE DISH. Top with pats of butter or margarine. Bake in a 350° oven for about 20 minutes.

SPREAD SOUR CREAM OVER THE POTATOES AND ADD THE SHREDDED CHEESE. Lower oven temperature to 325° and bake an additional 10 to 15 minutes or until bubbly and lightly browned on top.

Mashed Potatoes with Onion and Sour Cream

Yield: 4 to 6 servings

INGREDIENTS:

6 russet potatoes, peeled and quartered

1 cup sour cream

2 Tbsp. butter or margarine

salt and pepper to taste

1/2 cup finely chopped green onion

COOK POTATOES IN BOILING WATER UNTIL VERY TENDER. Drain and mash. Add sour cream, butter or margarine and salt and pepper to taste. Mix thoroughly until smooth. Add a small amount of whole milk if mixture is too stiff. Stir in the green onion.

VARIATION:

Left-Over Potato Patties

BRING LEFT-OVER MASHED POTATOES TO ROOM TEMPERATURE. Add beaten eggs (1 or more) and a little flour. Mixture should be stiff enough to form into patties. Add salt and pepper to taste.

PRESS PATTIES INTO SEASONED BREAD CRUMBS. Sauté in a skillet with vegetable or olive oil over medium heat.

Potato-Broccoli Casserole

Yield: 6 to 8 servings

INGREDIENTS:

5 medium potatoes, peeled, halved and sliced

2 cups broccoli florets

salt and pepper to taste

1 10 oz. can cream of mushroom soup

1-1/2 cups milk

1 cup shredded cheddar cheese

1/2 cup bread crumbs

1/2 cup grated Parmesan cheese

1/2 stick margarine

COVER POTATOES WITH BOILING WATER. Let stand a few minutes, then drain.

COOK BROCCOLI IN BOILING WATER approximately 5 minutes—then drain.

PUT POTATO SLICES INTO A GREASED CASSEROLE DISH, and add salt and pepper. Then distribute cooked broccoli over the potatoes. Blend soup with the milk and pour over all. Top this with the cheddar cheese. Mix bread crumbs with the Parmesan cheese and spread over the cheddar cheese. Put pats of margarine on top and bake uncovered in a 325° oven for approximately 35 to 40 minutes.

POTATO-CHEESE CASSEROLE

Yield: 4 servings

INGREDIENTS:

2-1/2 cups sliced, cooked potatoes

1 10 oz. can cream of celery soup

1/2 cup chopped onion

1/2 cup shredded cheddar cheese

1 tsp. dill weed

1 tsp. salt

1/4 tsp. coarse ground black pepper

2 Tbsp. grated Parmesan cheese

COMBINE ALL INGREDIENTS IN A BOWL. Transfer to a greased casserole dish and bake in a 325° oven for about 25 minutes or until bubbly and slightly browned on top.

POTATO PANCAKES

Yield: 10 to 12 potato pancakes

INGREDIENTS:

2 1/2 cups peeled, grated potatoes

3 eggs, beaten

2 Tbsp. (or more) flour

1/4 cup onion, minced

1/2 tsp. salt

1/2 tsp. black pepper

parsley flakes (optional)

1/4 cup vegetable oil

GRATE POTATOES AND BLOT WITH PAPER TOWELS.

IN A BOWL, MIX THE GRATED POTATOES with eggs, flour, onion and seasonings. Spoon 3 inch round cakes in hot cooking oil over medium heat, turning once to brown each side. Place on paper towel-lined plate to absorb excess oil before serving.

Fancy Buffet Potatoes

Yield: 8 to 10 servings

INGREDIENTS:

6 large potatoes, peeled, cooked and mashed

2 eggs, beaten

3 to 4 Tbsp. flour

1/2 tsp. salt

1/4 tsp. black pepper

1/4 cup additional flour for coating

2 additional eggs, beaten

5 ounces slivered almonds, crushed

4 Tbsp. margarine

COMBINE POTATOES, 2 BEATEN EGGS, FLOUR, SALT, AND PEPPER.

With a tablespoon, form potato mixture into balls. Coat these with additional flour, then dip into additional beaten eggs and roll in crushed almonds.

HEAT MARGARINE IN A SKILLET. Add potato balls, a few at a time, and brown gently on all sides. Add more margarine to skillet if needed.

Fancy Buffet Potatoes are a delicious addition to any special dinner. They are wonderful served with beef, pork or any meat entrée.

Greek Potatoes

Yield: 4 to 6 servings

INGREDIENTS:

3 or 4 russet potatoes

1/4 cup fresh lemon juice

3 to 4 Tbsp. butter or margarine

1/2 tsp. dried oregano

lemon pepper seasoning

PEEL AND SLICE POTATOES AND COOK IN ENOUGH WATER TO COVER in a skillet on top of the stove. When potatoes are nearly tender and water has almost cooked away, add the lemon juice, dried oregano and butter or margarine. Shake on lemon pepper seasoning to taste and cook on low a few more minutes.

Pizza Potatoes

Yield: 4 to 6 servings

INGREDIENTS:

4 large russet potatoes

1 cup prepared pizza sauce

1 to 2 Tbsp. butter or margarine

salt and pepper to taste

1-1/2 tsp. Italian seasoning

3/4 cup shredded Parmesan cheese

PEEL AND SLICE POTATOES AND COOK IN ENOUGH WATER TO COVER in a skillet on top of the stove. When potatoes are nearly tender and water has almost cooked away, add pizza sauce and butter or margarine. Add salt and pepper to taste and Italian seasoning, and cook a few more minutes. Top with Parmesan cheese and serve.

Twice-Baked Potato Casserole

Yield: 6 to 8 servings

INGREDIENTS:

6 large russet potatoes, baked

1/2 cup finely minced onion

1/2 pound hickory smoked bacon, cooked and crumbled

salt and pepper to taste

24 oz. sour cream

2 cups mozzarella cheese, shredded

2 cups cheddar cheese, shredded

1/2 cup finely chopped green pepper

1/2 cup finely chopped red pepper

1/2 cup sliced green onion

PEEL POTATOES, OR LEAVE SKINS ON IF PREFERRED. Slice potatoes and layer half in a greased 9x13 baking dish. Add 1/4 cup minced onion, salt and pepper, half of the bacon, 1/4 cup red pepper and 1/4 cup green pepper. Top with half of the sour cream and 1 cup of each shredded cheese.

REPEAT LAYERS AND BAKE UNCOVERED at 325° to 350° for about 20 minutes or until cheese is melted. Remove from oven and top with green onion.

RICE

When my children were growing up, the rice dish most familiar in our home was Rice-A-Roni. We also enjoyed Curried Rice, a side dish flavored with butter and curry powder.

My Spanish Rice recipe is easy to prepare, and is great served with Mexican food. When I first visited New Orleans, I discovered Dirty Rice, also known as Cajun Rice, so named because of the cooked chicken livers which give the dish a darkened appearance.

Rice and Pineapple is a nice addition to a meal featuring a pork roast or baked ham, and Wild Rice Dressing is a delicious alternative to the traditional bread dressing for poultry.

CURRIED RICE

Yield: 4 to 6 servings

INGREDIENTS:

2 cups Minute Rice

2 cups water

3 Tbsp. butter or margarine

1/2 tsp. salt

1/4 tsp. black pepper

2 tsp. curry powder

2 Tbsp. finely chopped fresh parsley

COOK RICE IN BOILING WATER according to package directions. Add butter or margarine and seasonings. Mix in the parsley, fluff with a fork and serve.

'Dirty' Rice

Yield: about 6 servings

INGREDIENTS:

4 oz. fresh chicken livers

3 Tbsp. butter or margarine

2 cups long grain white rice

4 cups chicken broth

1/4 tsp. crushed red pepper

1/4 tsp. salt

1/4 tsp. black pepper

1/4 tsp. garlic powder

1/4 tsp. ground thyme

1 2 oz. jar diced pimento, drained

1/2 cup finely chopped green onion

fresh parsley, minced

SAUTÉ LIVERS IN BUTTER OR MARGARINE for 5 to 8 minutes, or until brown on the outside and slightly pink on the inside. Remove from pan, cool and chop. Reserve pan drippings.

COOK RICE IN THE CHICKEN BROTH about 15 to 20 minutes or until chicken broth is absorbed. Add pan drippings, chopped chicken livers, seasonings, pimento, green onion and fresh parsley.

Quick Spanish Rice

Yield: 4 to 6 servings

INGREDIENTS:

2 cups Minute Rice

1-1/2 cups water

1 6 oz. can tomato juice

3 Tbsp. butter or margarine

1/4 tsp. ground cumin

3/4 cup salsa or picante sauce, mild or medium

1 Tbsp. fresh cilantro

BRING THE WATER AND TOMATO JUICE TO A BOIL. Add rice, cover and cook until the water is absorbed. Remove from heat. Let stand a few minutes and add butter or margarine, cumin, and salsa or picante sauce. Garnish with chopped cilantro.

Rice and Pineapple

Yield: 6 to 8 servings

INGREDIENTS:

2-1/2 cups Minute Rice

2-1/4 cups water

1 15 oz. can crushed pineapple, drained and juice reserved

6 Tbsp. brown sugar

butter

1/2 tsp. salt

1/4 tsp. curry powder

BRING WATER TO A BOIL IN SAUCEPAN AND ADD SALT. Slowly add rice and stir. Cover, remove from heat and let stand 3 to 5 minutes.

BUTTER BOTTOM OF GLASS BAKING DISH and put 1/3 of the cooked rice in the dish. Next add 1/2 the pineapple and top with pats of butter and 2 tablespoons brown sugar. Add another layer of rice, the remaining pineapple and more butter and 2 Tbsp. brown sugar. Top with remaining rice, a few more pats of butter and curry powder. Mix pineapple juice with 2 Tbsp. brown sugar and pour over all. Bake covered, at 325° for 20 to 30 minutes.

WILD RICE DRESSING

Yield: about 8 servings

Try this Wild Rice Dressing recipe with roast duckling or baked chicken. The recipe makes enough dressing for two large chickens or two 5 to 6 pound ducks.

INGREDIENTS:

BEGIN WITH:

2 boxes Uncle Ben's Wild Rice Mix with seasoning

2 cups fresh apple, peeled and chopped

1/3 cup finely chopped onion

1/2 cup raisins

1/2 tsp. sage

1/2 tsp. black pepper

1/4 cup melted butter or margarine

1-1/2 cups coarse bread crumbs

PREPARE WILD RICE MIX AS DIRECTED ON THE PACKAGE WITH THIS EXCEPTION: Five to ten minutes before the end of cooking time, add the apple, onion, raisins, sage, black pepper, margarine, and bread crumbs.

ALLOW TO SIMMER A FEW MINUTES ADDING A LITTLE WATER IF NEEDED— remove from heat and cool. Stuff cavity of poultry with the dressing (neck cavity also) and sew up openings or secure with skewers or wooden picks.

PLACE POULTRY IN LIGHTLY GREASED ROASTER PAN and bake in a 325° oven until temperature on meat thermometer reads about 165° to 170°.

DRESSING MAY ALSO BE BAKED IN A GREASED CASSEROLE DISH. Dot with butter or margarine and bake approximately 30 minutes in a 325° oven. Cover with foil for a portion of baking time for a more moist dressing.

Vegetables

    I remember well the wonderful fresh vegetables from my father's summer garden—among them cucumbers, fresh tomatoes and sweet corn. Dad began his work before planting season, with seeds in a small backyard hot house. Seedlings and plants were then transferred to a garden plot after the last frost in the Spring.

    My mother-in-law was also a gardener. Her vegetable offerings were delicious and were usually served in a rich cream sauce. My Sweet Potatoes and Oranges recipe is very similar to one she prepared for Holiday dinners.

    Vegetables, whether fresh, frozen or canned were always an important part of our family meals. There are several favorites we enjoyed every day, and others we prepared for special occasions.

BAKED ASPARAGUS

Yield: 6 servings

INGREDIENTS:

2 cups fresh asparagus, cut into 2 inch chunks, cooked and drained

2 hard-boiled eggs, chopped

2 10 oz. cans cream of mushroom soup

1/2 cup butter or margarine

1 tsp. salt

1/2 tsp. black pepper

1 cup cracker crumbs

1/2 cup grated Parmesan cheese

IN A 2-QUART GLASS BAKING DISH, ALTERNATE LAYERS OF ASPARAGUS, chopped egg and mushroom soup. Dot each layer with pats of butter, and season with salt and pepper. Top with cracker crumbs, pats of butter and Parmesan cheese. Bake in a 325° oven for 25 to 30 minutes.

Broccoli-Rice Casserole

Yield: 6 servings

INGREDIENTS:

2 cups broccoli florets

1 cup Minute Rice

2 Tbsp. butter or margarine

1 8 oz. pkg. Velveeta cheese

1 10 oz. can cream of mushroom soup

1/4 cup milk

1/2 tsp. coarse ground black pepper

Lawry's seasoned salt to taste

3/4 cup shredded cheddar cheese

COOK BROCCOLI IN WATER UNTIL JUST TENDER—about 8 to 10 minutes. Drain and set aside.

COOK RICE according to package directions. In a bowl, combine hot rice with broccoli and butter or margarine. Cut Velveeta cheese in small chunks and blend into the warm mixture. Add the soup and the milk and seasonings. Spoon mixture into a 2-quart casserole dish which has been sprayed with a vegetable coating. Top with shredded cheddar cheese.

BAKE IN A 300° OVEN ABOUT 20 MINUTES. Remove casserole from oven and allow to stand a few minutes before serving.

Cabbage and Noodles

Yield: 6 to 8 servings

INGREDIENTS:

1 large head cabbage, coarsely chopped

1 large onion, chopped

4 Tbsp. margarine

1 tsp. caraway seed

salt and pepper to taste

12 ozs. kluski noodles (or wide egg noodles)

IN A LARGE SKILLET, COOK CABBAGE AND ONION in margarine over low to medium heat. Add caraway seed and a little water and continue to cook until just tender. Add salt and pepper and set aside.

COOK NOODLES IN BOILING WATER AND DRAIN. Add these to the cabbage mixture and serve.

Baked Corn Casserole

Yield: 8 to 10 servings

INGREDIENTS:

2 8 oz. pkgs. corn bread mix

1 15 oz. can whole kernel corn, with juice

1 15 oz. can creamed corn

3/4 cup sour cream

1 egg, beaten

1/4 cup finely chopped green pepper

1/4 cup finely chopped red pepper

1/4 cup finely minced onion

parsley flakes to taste

salt and pepper to taste

2 Tbsp. margarine

IN A LARGE BOWL, COMBINE CORN BREAD MIX WITH ALL INGREDIENTS EXCEPT MARGARINE.

TURN MIXTURE INTO A GREASED 9x13 RECTANGULAR BAKING PAN or casserole dish and bake at 325° to 350° for approximately 40 to 50 minutes.

Melt 2 tablespoons of butter or margarine and drizzle over the corn casserole. Return to oven to bake an additional 10 minutes or until knife comes out clean.

Breaded Eggplant

Yield: 4 servings

INGREDIENTS:

1 large eggplant or 2 small

2 eggs, beaten

1/2 cup milk

1/2 cup flour

1 cup seasoned bread crumbs

1/2 cup additional flour

1/4 to 1/2 cup vegetable oil

PEEL EGGPLANT AND CUT INTO 1/2 INCH ROUNDS.

In a small bowl, mix together the eggs and milk. Coat eggplant slices with flour, dip them into the egg and milk batter, then into a mixture of the bread crumbs and 1/2 cup additional flour. Sauté eggplant in a skillet with vegetable oil, turning slices to brown.

VARIATION:
EGGPLANT PARMESAN

PREPARE EGGPLANT ACCORDING TO ABOVE RECIPE. Place cooked breaded eggplant in a greased glass casserole dish and spoon on marinara or spaghetti sauce. Bake in a 325° oven for about 15 minutes. Top with shredded Parmesan or mozzarella cheese and return to oven a few more minutes.

Sweet-Sour Green Beans

Yield: 6 to 8 servings

INGREDIENTS:

2 pounds or more fresh green beans

2 cups boneless smoked pork butt, diced

1 medium onion, chopped

coarse ground black pepper to taste

3 Tbsp. cornstarch (mixed with a little cold water)

3 Tbsp. white vinegar

2 Tbsp. granulated sugar

RINSE THE GREEN BEANS AND TRIM ENDS. Cut them or leave whole and put them in a kettle or large pan with water to cover. Add the smoked pork butt, onion and black pepper.

COOK OVER LOW HEAT approximately 30 to 45 minutes, or until beans are tender. Remove some of the hot liquid and whisk in the cornstarch, vinegar and sugar.

RETURN THIS MIXTURE TO THE GREEN BEANS and allow to simmer on low for 10 or 15 more minutes. Adjust vinegar and sugar to taste.

German Style Kraut

Yield: 4 servings

INGREDIENTS:

1 medium onion, chopped

3 Tbsp. butter

1 27 oz. can sauerkraut, drained

2 large potatoes, peeled and grated

8 oz. summer sausage, diced

1 15 oz. can chicken broth

2 Tbsp. or more brown sugar

1 bay leaf

coarse ground black pepper to taste

SAUTÉ ONION IN BUTTER IN A DEEP SKILLET, and when transparent, add the sauerkraut. Add potatoes, summer sausage, broth, brown sugar, bay leaf, and black pepper. Simmer mixture about 30 minutes. Stir once or twice and remove bay leaf before serving.

Baked Squash

Yield: 2 servings

INGREDIENTS:

1 acorn or butternut squash

2 Tbsp. butter or margarine

2 Tbsp. brown sugar

coarse ground black pepper to taste

WITH A SHARP KNIFE, CUT SQUASH INTO HALVES and remove pulp and seeds. Place cut side down in a baking pan. Add a little water and bake for approximately 20 minutes in a 325° to 350° oven. Remove pan from oven and turn squash cut side up. Into each cavity spread the butter or margarine and brown sugar. Shake on a little coarse ground black pepper. Return to oven for another 20 to 30 minutes or until squash is fork tender.

Sauerkraut With Bacon

Yield: 4 to 6 servings

INGREDIENTS:

6 slices hickory smoked bacon, diced

1 medium onion, chopped

2 16 oz. cans sauerkraut

1 tsp. caraway seed

2 or 3 Tbsp. brown sugar

salt and pepper to taste

1 Tbsp. cornstarch mixed with 1 cup of water

FRY BACON AND ONION IN A SKILLET ON TOP OF THE STOVE. In a saucepan, combine the sauerkraut, caraway seed, brown sugar and salt and pepper. Add the onion and bacon with drippings to the sauerkraut mixture. Cover pan and simmer on low for about 15 to 20 minutes. Mix cornstarch with water and add to the kraut mixture to thicken. Simmer a few more minutes before serving.

Sweet Potatoes and Oranges

Yield: 6 to 8 servings

INGREDIENTS:

4 or 5 medium fresh sweet potatoes, cooked, peeled and sliced

1 or 2 oranges, halved and thinly sliced (rind left on and seeds removed)

2 Tbsp. brown sugar

4 Tbsp. butter

1/2 cup orange juice

nutmeg

cinnamon

LAYER SWEET POTATOES AND ORANGES IN A LIGHTLY GREASED GLASS BAKING DISH. Top each layer with brown sugar and butter and pour on the orange juice. Place dish in a 325° oven and bake until hot—about 20 minutes.

REMOVE FROM OVEN and top with more brown sugar and butter if desired. A dash of nutmeg or cinnamon mixed with the brown sugar is very good.

Favorite Sweet Potatoes

Yield: 8 to 10 servings

INGREDIENTS:

6 medium to large sweet potatoes

coarse ground black pepper to taste

1/2 stick or more butter or margarine

4 Tbsp. honey

1/2 cup brown sugar

COOK THE SWEET POTATOES IN WATER UNTIL NEARLY DONE. Cool and peel them.

IN A GREASED GLASS CASSEROLE DISH, layer slices of sweet potatoes—season with a little black pepper. Top each layer with pats of butter or margarine, then add honey and brown sugar as desired.

BAKE IN A 325° OVEN ABOUT 20 MINUTES or until heated through and golden brown.

VEGETABLE STIR-FRY

Yield: 6 to 8 servings

INGREDIENTS:

2 small yellow (summer) squash, rinsed and sliced

2 small or medium zucchini squash, rinsed and sliced

1 medium onion, sliced lengthwise

2 red bell peppers, seeded and julienne sliced

2 green bell peppers, seeded and cut in chunks

3 Tbsp. butter or margarine

coarse ground black pepper to taste

1-1/2 tsp. garlic powder

1 tsp. Italian seasoning

SAUTÉ SQUASH, ONION AND PEPPERS in large skillet with butter or margarine until just tender. Continue stirring while vegetables cook and add seasonings as desired.

Zucchini and Tomatoes

Yield: 4 to 6 servings

I first tasted this delicious vegetable dish during a summer trip to a resort in Minnesota. The individual who introduced me to this dish was an expert gardener and prepared it often during the growing season.

INGREDIENTS:

4 to 5 cups sliced zucchini squash

1 cup onion, chopped or sliced lengthwise

3 Tbsp. butter or margarine

1 14 oz. can stewed or diced tomatoes

2 or 3 cloves of fresh garlic, chopped

salt and pepper to taste

SAUTÉ THE ZUCCHINI AND ONION in butter or margarine in a large saucepan on top of the stove.

ADD STEWED TOMATOES, garlic and salt and pepper. Cook on low for about 20 to 30 minutes, and add a little water if needed. Stir once or twice while cooking.

FOR A LOW CALORIE MEAL, thicken the mixture with a small can of tomato sauce, add a pinch of Italian seasoning and serve over cooked angel-hair pasta.

ZUCCHINI-POTATO PATTIES

Yield: about 8 vegetable patties

INGREDIENTS:

3 potatoes, peeled and shredded

2 cups fresh zucchini, shredded

2 eggs, beaten

1 cup bread crumbs

1/2 cup grated Parmesan cheese

salt and pepper to taste

SQUEEZE EXCESS MOISTURE FROM SHREDDED POTATOES AND ZUCCHINI. Add eggs, bread crumbs, Parmesan cheese and seasoning. Mixture should be firm enough to form into patties. Sauté in vegetable oil until brown on both sides.

Desserts

    Although I have never been an expert baker, I have prepared my share of desserts over the years; however, everyday offerings for my family were usually canned fruit, Jell-O or pudding, and now and then cupcakes or brownies.

    I'm sharing a few family favorites, such as Homemade Pumpkin Pie, which calls for fresh cooked pumpkin, rather than canned, and has a wonderful flavor and texture. The Christmas Spumoni Cookies will be a nice addition to a Holiday dessert table.

    Lemon Meringue Pie was one of my mother's specialties, and, although I regret not having her exact recipe, my lemon pie comes close. The Fruit Compote, which my stepmother often served, was always a light and delicious way to end a meal.

Hawaiian Cake

Yield: about 12 servings

INGREDIENTS:

1 pkg. yellow cake mix

2 cups cold milk

2 3 oz. pkgs. instant vanilla pudding mix

1 8 oz. pkg. Philadelphia cream cheese, softened

1 8 oz. carton Cool Whip

1 20 oz. can crushed pineapple, drained

1/2 cup maraschino cherries, drained, and chopped

1 cup flaked coconut

1/2 cup macadamia nuts, chopped

PREHEAT OVEN TO 350°.

PREPARE CAKE MIX ACCORDING TO PACKAGE DIRECTIONS and pour batter into a greased 9x13 baking pan. Bake at 350° about 30 to 35 minutes. Cool completely.

IN A MEDIUM SIZED BOWL, combine the milk, pudding mix and cream cheese.

Fold in the Cool Whip and spread on the cake.

TOP WITH pineapple, cherries, coconut and nuts.

REFRIGERATE UNTIL SERVING TIME.

Pineapple-Orange Upside-Down Cake

Yield: about 6 servings

INGREDIENTS:

1-1/2 cups Bisquick baking mix

1/2 cup granulated sugar

1/2 cup milk

3 Tbsp. vegetable oil

1 tsp. vanilla or orange extract

2 eggs, beaten

1/4 cup butter or margarine

1/2 cup brown sugar

1 cup canned pineapple chunks, juice drained

1 11 oz. can Mandarin oranges, juice drained

maraschino cherries

whipped topping

MIX TOGETHER the Bisquick, sugar, milk, oil, vanilla or orange extract and eggs.

SPRAY A 9-inch pie pan with a vegetable coating spray. Put several pats of butter or margarine in pan, and top with brown sugar. Arrange pineapple and oranges over this and add a few maraschino cherries.

NEXT pour the mixed batter over the fruit. Bake in a 350° oven for about 35 minutes, or until a toothpick inserted in the cake comes out clean. Invert the cake at once onto a platter. Serve with whipped topping.

Pumpkin Cake

Yield: about 12 servings

INGREDIENTS:

1 box yellow cake mix

3/4 cup butter or margarine, softened

1 30 oz. can Libby's Pumpkin Pie Mix

2 eggs, beaten

1 12 oz. can evaporated milk

PREHEAT OVEN TO 350°.

COMBINE CAKE MIX with the butter or margarine and mix until crumbly. Measure one cup of the mixture and set aside. Press remaining mixture into a greased 9x13 baking pan.

COMBINE the pumpkin pie mix with the eggs and evaporated milk. Blend until smooth.

POUR MIXTURE INTO THE PAN over the cake mix base and top with the reserved cup of cake mix.

BAKE IN A 350° oven for approximately 30 to 40 minutes.

Lemon 'Cheesecake'

Yield: about 12 servings

INGREDIENTS:

GRAHAM CRACKER CRUST:

30 graham crackers

2 tsp. granulated sugar

1 stick margarine, melted

FILLING:

2 3 oz. pkgs. regular lemon gelatin

2 cups boiling water

2 3 oz. pkgs. Philadelphia cream cheese, softened

1 cup granulated sugar

2 cups Cool Whip or whipped topping

CRUSH GRAHAM CRACKERS, add sugar and margarine and mix well. Put about two thirds of this crust mixture into a 9x13 glass baking dish and pat down. Remainder is reserved for topping.

DISSOLVE THE GELATIN IN BOILING WATER and chill until partially thickened. Blend together the cream cheese and sugar.

COMBINE THE GELATIN WITH CREAM CHEESE MIXTURE and fold in the whipped topping. Spread mixture over the graham cracker crust and chill at least 4 hours. Top with remaining graham cracker mixture.

Note:
This Lemon Cheesecake may be put in the freezer for a few minutes before serving for a more firm dessert.

Christmas 'Spumoni' Cookies

Yield: About 4 dozen cookies

INGREDIENTS:

1 cup butter, softened

2 cups granulated sugar

2 eggs, beaten

3 cups all-purpose flour

1/2 tsp. salt

3 tsp. double-acting baking powder

red food coloring paste

green food coloring paste

1/2 tsp. peppermint extract

1/2 tsp. almond extract

1/2 tsp. vanilla extract

1/2 cup candied red cherries, chopped

1/2 cup candied green cherries, chopped

1/2 cup English walnuts, finely chopped

MIX BUTTER WITH GRANULATED SUGAR—blend in the beaten eggs.

Sift flour with salt and baking powder and stir into the butter-sugar-egg mixture.

DIVIDE DOUGH INTO THIRDS.

To the first portion of dough, add red food coloring paste, peppermint extract, and red candied cherries.

To the next portion of dough, add the green food coloring paste, almond extract and green candied cherries.

To the remaining portion of dough, add vanilla extract and finely chopped English walnuts.

CHILL EACH PORTION OF DOUGH SLIGHTLY, then roll (onto individual sheets of waxed paper) into three rectangles about 1/8 inch thick. Chill each several hours then stack rectangles and roll together like a jelly roll. Chill again until firm, then slice and bake on a greased and floured cookie sheet in a 350° oven for 8 to 10 minutes. Bake a few 'test' cookies to determine the best oven temperature and baking time.

Note: Add enough food coloring paste to reach desired red or green color.

The portion of dough without food coloring paste should be the center cookie layer.

Fruit Compote

Yield: 4 to 6 servings

INGREDIENTS:

3 to 4 fresh oranges, peeled and sectioned

1 pint fresh strawberries, cored and sliced

4 sliced bananas

1 10 oz. pkg. frozen sliced strawberries, thawed

COMBINE INGREDIENTS IN A LARGE BOWL.

MIX GENTLY AND ALLOW FLAVORS TO BLEND. Refrigerate before serving.

APPLE PIE

Yield: 6 to 8 servings

INGREDIENTS:

1 pkg. refrigerated pie crust

6 cups peeled and sliced apples (Jonathan or Granny Smith apples are best)

2 Tbsp. flour

3/4 cup granulated sugar

1/2 tsp. ground cinnamon

1/2 tsp. ground nutmeg

butter or margarine

IN A BOWL, MIX SLICED APPLES, flour, sugar, cinnamon and nutmeg. Let stand a few minutes to blend flavors.

SPRAY PIE PLATE WITH VEGETABLE COATING SPRAY and dust with flour. Line pie plate with prepared bottom crust and fill with fruit mixture. Top with a few pats of butter or margarine. Cover apple filling with top crust, crimp edges to seal, and cut several small slits in crust. Bake in a 350° to 375° oven for about 45 to 50 minutes.

TO PREVENT CRUST FROM BROWNING TOO QUICKLY, form strips of foil around the edges of pie pan and secure.

**VARIATION:
LESS-SUGAR APPLE PIE**

Follow the above recipe for Apple Pie with this exception: use 1/2 cup granulated sweetener (for baking) and 1/4 cup granulated sugar.

Lemon Meringue Pie

Yield: 6 to 8 servings

INGREDIENTS:

1 cup granulated sugar
1/2 cup cornstarch
1-1/2 cups cold water
1/4 cup lemon juice
1 tsp. lemon extract
3 egg yolks, beaten
1 Tbsp. melted butter or margarine
1 Tbsp. lemon zest

MIX TOGETHER IN A SAUCEPAN the sugar and cornstarch and add the cold water, lemon juice and lemon extract. Cook on medium heat, stirring constantly until mixture comes to a full boil.

REMOVE FROM HEAT. Place about 1 cup of the hot mixture in a small bowl, and mix in the beaten egg yolks. Add this back into the saucepan, return to heat and cook another minute or two until thick and bubbly. Add butter or margarine and lemon zest.

POUR INTO A 9-INCH BAKED PIE SHELL and prepare meringue.

MERINGUE:

3 egg whites
4 Tbsp. granulated sugar
1/2 tsp. vanilla

BEAT EGG WHITES UNTIL FOAMY— gradually add sugar and vanilla extract. Continue to beat until stiff peaks form.

SPREAD MERINGUE OVER PIE WHILE LEMON FILLING IS STILL WARM, and secure meringue to edges of the pie crust. Bake 10 to 15 minutes at 325° or until peaks are browned.

Homemade Pumpkin Pie
(Made with Fresh Cooked Pumpkin)

Yield: 6 to 8 servings per pie

CUT SMALL PUMPKINS OR 'PIE' PUMPKINS IN HALF. Remove seeds and stringy pulp. Put pumpkins in a baking pan, cut side down with a little water added and bake in a 325° to 350° oven for about an hour, or until pumpkins are tender. Allow to cool, and peel away shells. Cut into sections and process in food processor or blender. Place the pumpkin in a colander for about an hour to drain excess liquid.

INGREDIENTS:

FOR TWO PUMPKIN PIES, YOU WILL NEED:

- 2-1/2 cups cooked pumpkin
- 1-1/2 cups granulated sugar
- 1 tsp. salt
- 2-1/2 tsp. ground cinnamon
- 1 tsp. ground ginger
- 1/2 tsp. ground cloves
- 1 tsp. ground nutmeg
- 5 beaten eggs
- 2 12 oz. cans evaporated milk
- pie crust for two 9-inch deep-dish pies

COMBINE IN A LARGE BOWL the sugar, salt, cinnamon, ginger, cloves and nutmeg. Gradually add the eggs, milk and pumpkin to the sugar and spices. Blend well. Pour mixture into two unbaked pie shells and bake in a 375° oven for about 12 minutes. Lower oven temperature to 325° and bake about 45 minutes more or until knife inserted into pie comes out clean.

TO PREVENT CRUSTS FROM BROWNING TOO QUICKLY, form strips of foil around the edges of pie pans and secure.

Note:
Spices may be adjusted as preferred, and brown sugar may be substituted for a portion of the total amount of sugar.

Pumpkin Mincemeat Squares

Yield: 12 servings

INGREDIENTS:

3/4 cup margarine, softened

1 cup granulated sugar

1 cup brown sugar

4 beaten eggs

2-2/3 cups sifted flour

2-1/2 tsp. baking powder

1 tsp. ground cinnamon

1 tsp. ground nutmeg

1 15 oz. can pumpkin

1 15 oz. jar mincemeat

1 cup chopped pecans

COMBINE MARGARINE with granulated sugar and brown sugar, and mix well. Add beaten eggs.

SIFT TOGETHER THE FLOUR, BAKING POWDER, cinnamon, and nutmeg. Add alternately the pumpkin and mincemeat. Blend well after each addition.

STIR IN THE CHOPPED PECANS and bake in a 9x13 greased and floured baking pan at 325° for 35 to 45 minutes.

When cool, frost with cream cheese frosting and cut into squares.

CREAM CHEESE FROSTING:

1/2 cup butter, room temperature

8 oz. pkg. Philadelphia cream cheese, room temperature

2 to 3 cups powdered sugar

1 tsp. vanilla extract

WITH AN ELECTRIC MIXER, blend the butter and cream cheese together, about 3 minutes on medium speed until very smooth.

ADD AND MIX IN THE VANILLA EXTRACT. Add powdered sugar slowly until desired thickness is reached.

SPREAD FROSTING on bars. Cut and serve.

Strawberry Pie

YIELD: 6 to 8 slices

INGREDIENTS:

baked pie crust

4 to 6 cups sliced fresh strawberries

1 cup granulated sugar

2 Tbsp. cornstarch

1 cup water

1 3 oz. pkg. strawberry or raspberry gelatin

a few drops of red food coloring (optional)

COMBINE SUGAR AND CORNSTARCH IN SAUCEPAN ON TOP OF THE STOVE. Add water and bring to a boil. Lower heat and stir constantly until mixture thickens—about 2 minutes. Remove from heat and add the package of gelatin, stirring to dissolve. Cool the mixture slightly before adding the sliced strawberries and pour into a baked pie shell. Chill until firm.

Top with whipped cream and whole strawberries.

SAUCE AND GRAVY

    The purpose of a sauce is to season and enhance the flavor of a particular food. A sauce can be savory, spicy or sweet, to complement meats, vegetables or desserts. I have included my suggestions for making gravy, Basic White Sauce and Homemade Cocktail Sauce as well as two versions of Homemade Spaghetti Sauce.

    The first time I learned that cornstarch could be used to thicken gravy was many years ago when a friend invited us to her home for dinner. She prepared a beef pot roast and thickened the pan drippings with cornstarch. The gravy was delicious and of the right consistency. I have been a fan ever since.

All Purpose Gravy

INGREDIENTS:

pan drippings
flour or cornstarch
Kitchen Bouquet

GRAVY IS MADE WITH PAN DRIPPINGS OR MEAT JUICES, whether from beef, pork, poultry, lamb, or veal.

GRAVY MAY BE MADE in a skillet on top of the stove, for instance when making fried chicken or pork chops. After frying, transfer meat to a platter and discard excess fat from the pan, with an ample amount of drippings and crusty bits remaining. Over low heat, add about 1/2 cup of flour to the drippings and stir with a whisk. Add milk or water to desired thickness as gravy slowly simmers—add salt and pepper to taste.

FOR OVEN ROASTED MEATS, I RECOMMEND THE FOLLOWING:

COLLECT PAN JUICES FROM ROASTER PAN WHILE THE MEAT IS STILL COOKING. Add small amounts of water to the roaster as you remove the meat juices—this will increase the amount of pan drippings for gravy. Transfer pan juices to a saucepan on top of the stove.

Drippings may be put in the freezer for 20 to 30 minutes. The fat comes to the top and can be easily discarded.

THE GENERAL RULE IS:

For every two tablespoons of pan drippings or meat juices, use about 2 tablespoons of flour or cornstarch and one cup of water or broth. Mix flour or cornstarch with a little cold water before adding to the drippings.

All Purpose Gravy (Continued)

A small amount of Kitchen Bouquet is a great 'gravy partner'- adding both color and flavor. Bouillon cubes and packaged gravy mixes can also be used. For a beef entrée, one of the best mixes is McCormick Onion Gravy Mix. Salt, pepper, or other seasonings may be added to gravy as desired.

Fresh mushrooms, sliced and sautéed in butter or margarine, are a nice addition to gravy.

Homemade Cocktail Sauce For Seafood

Yield: about 2 cups

INGREDIENTS:

1 12 oz. bottle of chili sauce

2 Tbsp. prepared fresh ground horseradish

1 Tbsp. finely minced onion

1/4 cup sweet pickle relish

COMBINE ALL INGREDIENTS AND MIX WELL, adjusting additions to the cocktail sauce as preferred.

Great with boiled shrimp and deep fried shrimp.

Sour Cream-Dill Gravy

Yield: 3 cups

INGREDIENTS:

1/2 cup pan drippings

1 15 oz. can chicken broth

1 tsp. dill weed

1/4 tsp. black pepper

1 cup sour cream

ADD TO PAN DRIPPINGS the can of chicken broth and stir over medium-high heat. Mix in dill and black pepper, and turn heat to low.

TO THICKEN GRAVY, a small amount of cornstarch mixed with cold water may be added. Add sour cream and stir until smooth.

Crock Pot Spaghetti Sauce

Yield: about 12 cups

INGREDIENTS:

6 links mild Italian sausage

1 pound or more pork neck bones*

2 15 oz. cans tomato sauce

3 12 oz. cans tomato paste

3 cups water

3/4 cup chopped onion

1 tsp. instant minced garlic

3 Tbsp. granulated sugar

2 Tbsp. Italian seasoning

1-1/2 tsp. ground oregano or oregano leaves

1 Tbsp. coarse ground black pepper

1 tsp. salt

2 Tbsp. parsley flakes

2 4.5 oz. jars sliced mushrooms, drained

MIX the tomato sauce, tomato paste, water, onion, and all seasonings into the crock pot. Add the mushrooms, Italian sausage links and pork neck bones. Cook on LOW about 6 to 8 hours, stirring once or twice if possible.

REMOVE AND TRIM MEAT FROM NECKBONES and return to the sauce. Carefully remove the Italian sausage and keep warm. Stir the sauce, and if not desired thickness, add a small can of tomato paste. If sauce is too thick, add a little water or tomato juice. Stir again, and cook an additional 10 to 15 minutes. Serve the Italian sausage and sauce with any type of cooked pasta.

*Pork neck bones can be added to crock pot in a type of bouquet–garni bag, made from a large piece of cheesecloth, and tied with string or butcher twine. This will prevent the possibility of small pieces of bone in the sauce.

Homemade Spaghetti Sauce

Yield: about 12 cups

INGREDIENTS:

6 links mild Italian sausage

3 12 oz. cans tomato paste

2 15 oz. cans tomato sauce

5 to 6 cups water

2 Tbsp. finely chopped garlic

1/2 cup diced onion

1 tsp. sweet basil

1-1/2 tsp. Italian seasoning

1 tsp. ground oregano

2 Tbsp. granulated sugar

salt and pepper to taste

IN A LARGE KETTLE OR STOCK POT, combine tomato paste, tomato sauce, water, garlic, onion, sugar, and all seasonings. Add the links of Italian sausage and simmer on low for 3 to 4 hours, stirring occasionally.

SERVE SAUCE AND ITALIAN SAUSAGE OVER COOKED SPAGHETTI OR MOSTACCIOLI.

Note:
A package of dry spaghetti sauce mix may be added to either this recipe or to Crock Pot Spaghetti Sauce for extra flavor and thickening.

Homemade Tartar Sauce For Fish or Seafood

Yield: 1-1/2 cups

INGREDIENTS:

1 cup mayonnaise

1 tsp. prepared mustard

1 Tbsp. parsley flakes

1 Tbsp. finely minced onion

1 2 oz. jar diced pimento, drained

1 hard-boiled egg, chopped (optional)

COMBINE ALL INGREDIENTS. Season according to taste. A small amount of lemon juice, vinegar, horseradish, sweet pickle relish, or any combination of these may also be added. Serve with fried or baked fish.

SAUCES FOR MEATBALLS

INGREDIENTS:

TOMATO SAUCE:

3 cups catsup

1/2 cup cider vinegar

1/2 cup water

2 tsp. black pepper

1/2 to 1 tsp. dry mustard

2 tsp. chili powder

1/4 tsp. crushed red pepper

2 Tbsp. Worcestershire Sauce

2 Tbsp. cornstarch mixed with 1/4 cup cold water

CRANBERRY SAUCE:

1 can cranberry sauce

12 oz. chili sauce

TOMATO SAUCE:

COMBINE ALL INGREDIENTS IN SAUCEPAN AND HEAT UNTIL THICKENED. Pour over meatballs in crock pot or chafing dish.

CRANBERRY SAUCE:

Mix equal amounts of bottled chili sauce with canned whole or jellied cranberry sauce.

HEAT CRANBERRY SAUCE, pour over meatballs and keep warm in crock pot or chafing dish.

Basic White Sauce

Yield: about 1 cup

INGREDIENTS:

2 Tbsp. butter

2 Tbsp. flour

1 cup milk or half & half

salt and coarse black pepper to taste

MELT BUTTER IN SAUCEPAN ON TOP OF THE STOVE OVER MODERATE HEAT. Add flour and stir quickly—drizzle in 1 cup of milk or half & half.

Note:
Recipe can be doubled and is delicious with vegetables such as peas or carrots. Add shredded cheddar cheese for creamed cauliflower.

Horseradish Sauce

Yield: about 1 cup

INGREDIENTS:

1 cup Basic White Sauce (See page 198)

3 Tbsp. prepared horseradish

2 Tbsp. heavy cream

1 Tbsp. granulated sugar

1 tsp. prepared mustard

1 Tbsp. cider vinegar

COMBINE THE WHITE SAUCE with all other ingredients in order given. Stir well to blend.

Serve with roast beef or prime rib.

Food Safety Guidelines

CLEANLINESS

Safe steps in food handling, cooking and storage are essential to prevent food-borne illness. Follow these guidelines to keep food safe:

- Clean: Wash hands and surfaces often
- Separate: Don't cross-contaminate
- Cook: Cook to proper temperature
- Chill: Refrigerate promptly

THAWING

- It is best to thaw frozen meats and poultry in the refrigerator. Meat and poultry can be thawed in a plastic bag submerged in water and should be cooked immediately after thawing. Meat thawed in the microwave must also be cooked immediately after thawing.

PREPARATION

- Wash hands and don't cross-contaminate. Keep raw meat, poultry, fish and their juices away from other food. After cutting raw meats wash cutting board, utensils and surfaces with hot, soapy water.
- Marinate meat and poultry in a covered dish in the refrigerator. Discard left-over marinade.

COOKING

- See recommended cooking instructions for meats on the following pages.

SERVING

- When serving food at a buffet, keep hot food hot with the use of chafing dishes, slow cookers or warming trays. Hot food should be held at a minimum temperature of 140°.

- Cold foods should be displayed in small dishes or trays and nested in bowls of ice water.
- Food should not be displayed more than two hours at room temperature or a maximum of one hour outdoors when the temperature is above 90°.

LEFTOVERS
- Food should be stored in the refrigerator and used within 4 days or should be frozen. Foods at a buffet which are left out longer than the recommended period of time should be discarded.

Cooking Instructions
FOR FRESH MEATS

BEEF

GROUND BEEF:
Cook or Grill to internal temperature of 160° or until meat juices run clear.

STEAK:
Grill or Broil to "Medium." Meat may be pink in the center, but must be well-seared on both sides.

PRIME RIB AND BONELESS BEEF ROAST:
Roast uncovered at 300° for one hour. Cover pan, lower oven temperature to 200° and continue roasting until internal temperature reads 145° (for medium) on meat thermometer.

PORK

FRESH PORK LOIN:
Roast uncovered at 325° for one hour. Cover pan, lower oven temperature to 275° and continue roasting until internal temperature reads 145° on meat thermometer.

LAMB

Roast or Grill until internal temperature registers 145° on meat thermometer. Lamb chops must be well-seared on both sides.

POULTRY

Roast or Grill until internal temperature reads 165° on meat thermometer.

The USDA recommends the above internal temperatures for Beef, Pork, Lamb and Poultry. For reasons of personal preference, pork and lamb may be cooked to 160° (medium.) Poultry must be cooked to a minimum internal temperature of 165°. Allow meat or poultry to rest at least 3 minutes before carving and serving.

Use of a meat thermometer is essential to assure that meat is cooked to a safe internal temperature in order to prevent food borne illness.

Cooking Instructions
for Smoked Meats and Sausage

SMOKED HAM:
Smoked Ham is fully cooked. Heat at 300° (15 to 20 minutes per pound) until internal temperature reads 140° on meat thermometer.

SMOKED TURKEY:
Smoked Turkey is delicious served cold. However, if heating is preferred, wrap in foil and heat in a 250° oven until internal temperature reads 140° on meat thermometer.

THREE CATEGORIES OF SAUSAGE

FRESH SAUSAGE:
Fresh Polish, Italian, Bratwurst, etc.
Fresh Sausage must be fully cooked.
Skillet: Simmer in water a total of 30 to 40 minutes. Turn to brown.
Oven: Bake at 325° for about one hour.
Grill: Grill 10 to 15 minutes per side, about 30 minutes total cooking time.

SMOKED SAUSAGE:
Smoked Polish, Hot Dogs, Mini-Polish, etc.
Smoked Sausage is fully cooked and ready to eat.
Heating is recommended for full flavor enjoyment.
Skillet: Heat in water for 10 to 15 minutes.
Oven: Bake at 325° for about 20 minutes.
Grill: Grill about 5 minutes per side.

COOKED SAUSAGE:
Bockwurst, Kiszka, or Jaternice/Jelita
Cooked Sausage items are fully cooked.
Heating is recommended for full flavor enjoyment.
Skillet: Simmer in water on low just until hot.

INDEX

1000 Island Dressing 45
Acorn squash 169
All Purpose Gravy 190
Appetizer Ribs 12
Apple Pie 184
Apple Sausage Pancake 2
Apricot Pork Pecan 134
Artichoke hearts 13
Artichoke Spinach Spread 13
Bacon Brunch Muffins 3
Bacon-Onion Rolls 117
Baked Asparagus 162
Baked Chicken With Stuffing 61
Baked Corn Casserole 165
Baked Italian Fries 142
Baked Squash 169
Barbeque Sauce 12
Barbequed Pork Back Ribs 96
Basic White Sauce 198
Beef and Cheese Burgers 113
Beef chuck roast 52
Beef liver 79
Beef Pot Roast 53
Beef Roast-Marinade 54
Beef shank meat 33
Beef Short Ribs 55
Beef stew meat 106

Beef Stroganoff 57
Beef Sweetbreads 58
Beer Bratwurst 112
Better Au Gratin Potatoes 143
Bone-in chicken breasts 29, 62, 116
Braised Oxtails 56
Bread Dumplings 138
Breaded Eggplant 166
Breaded Fish Filets 127
Breaded Veal Cutlets 108
Breaded-Minute Steak 102
Breads for the Grill, Two Ways ... 119
Breakfast Quiche 4
Broccoli florets 28, 89, 92, 147, 163
Broccoli-Cauliflower Cheese Soup ... 28
Broccoli-Rice Casserole 163
Brunch Casserole 5
Bulk pork sausage 2, 9, 20, 21,
    81, 82, 84, 85, 100, 134-138
Butternut squash 169
Cabbage and Noodles 164
Cabbage Slaw 38
Camper's Stew 124
Canadian bacon 5, 6
Cauliflower florets 28, 89, 198
Cheddar cheese 5, 7, 8, 23, 28, 39,
    67, 80, 98, 114, 145, 147, 153, 163, 198

Cherry Glazed Ham 75	Dirty Rice 157
Cherry tomatoes 14, 45, 48	Dried beef 16
Chicken 29, 59-66, 74, 115, 116, 118	Duck A L'Orange 71
Chicken A la King 59	Duckling 71
Chicken and Noodles 63	Easy Oven Dinner 72
Chicken Cacciatore 62	Easy Potato Dumplings 144
Chicken Fried Steak 101	Easy Swedish Sauce 20
Chicken livers 19, 157	Eggplant Parmesan 166
Chicken Paprikás (Paprika Chicken) 65	Eggs 6, 8, 18
Chicken Parmesan 66	Eggs-Benedict Canadian 6
Chicken Soup 29	Fancy Buffet Potatoes 150
Chili Con Carne 67	Favorite Sweet Potatoes 172
Chili-Nacho Dip 15	Favorite Vegetable Soup 31
Chipped Beef Dip 16	Filé powder 74, 125, 130
Chorizo Sausage and Rice 68	Fruit Compote 183
Christmas 'Spumoni' Cookies 182	Garlic Grill Loaf 119
Corn Dogs 69	German Pork Cutlets 90
Corned beef 22, 70, 94, 95	German Style Kraut 168
Corned Beef and Cabbage 70	Glazed Chicken On The Grill 115
Cottage cheese 39, 46	Granny Smith apples 2, 184
Cottage Cheese Salad 39	Great Northern Bean Soup 32
Crab 14, 126	Greek Potatoes 151
Crab and Shrimp Cakes 126	Green beans 167
Cranberry Sauce 40, 197	Grilled Rye Loaf 119
Cran-Raspberry Jell-O Salad 40	Ground beef 15, 20, 67, 73,
Cream Cheese Frosting 187	80, 81, 82, 84, 85, 104, 113, 114, 135, 203
Creamy Potato Soup 30	Ham and Cheddar Cups 7
Crock Pot Bean Soup 32	Hamburger Goulash 73
Crock Pot Beef Chuck Roast 52	Hash Brown and Egg Bake 8
Crock Pot Sausage and Cabbage 97	Hash Brown Potato Bake 145
Crock Pot Spaghetti Sauce 194	Hawaiian Cake 178
Cucumber Salad 41	Hearty Beef Soup 33
Cucumber Sandwiches 17	Hickory smoked bacon 3, 4, 7, 8,
Cucumbers and Sour Cream 42	..... 47, 48, 50, 110, 111, 113, 117, 153, 170
Curried Rice 156	Hickory smoked bone-in ham 75, 120
Deviled Eggs 18	Hickory smoked boneless ham 7
Dill pickles 22	Homemade Cocktail Sauce 192

Homemade Pumpkin Pie 186	*Oriental Pork Loin* 136
Homemade Spaghetti Sauce 195	*Oven Baked Stew* 106
Homemade Tartar Sauce 196	*Oven BBQ Chicken* 64
Horseradish Sauce 199	*Oxtail Soup* 34
Iceberg lettuce 45	Oxtails 34, 56
Italian Meatballs 81	*Philly Steak Sandwiches* 105
Italian sausage 23, 99, 194, 195	*Pickle Roll-Ups* 22
Jonathan apples 2, 49, 184	*Pineapple-Orange* ....
Kluski noodles 57, 164	*Upside-Down Cake* 179
Knockwurst and Kraut 76	*Pizza Potatoes* 152
Knockwurst sausage 76	*Polish Sausage Luau* 121
Lamb shanks 78	*Polish Style Pork Loin* 91
Left-Over Potato Patties 146	*Polynesian Grilled Chicken* 118
Leg of Lamb 77	*Pork and Dumpling Roast* 138
Lemon 'Cheesecake' 181	*Pork and Sweet-Sour Cabbage* 93
Lemon Chicken 60	Pork back ribs 96, 123
Lemon Meringue Pie 185	Pork Cheeks 86
Less-Sugar Apple Pie 184	Pork chops 87, 88, 93
Liver and Onions 79	*Pork Chops and Gravy* 87
Liver Spread 19	*Pork Chow Mein* 89
Macaroni Shell Salad 43	Pork loin back ribs 12
Mashed Potatoes With	*Pork Polynesian* 137
Onion and Sour Cream 146	*Pork Stir-Fry* 92
Meatball Sandwiches 81	Pork tenderloin 83, 92
Mexican Beef Roll 135	*Potato Pancakes* 149
Mexican Lasagna 80	*Potato-Broccoli Casserole* 147
Mexican Pork Stew (Posole) 83	*Potato-Cheese Casserole* 148
Microwave Fish Filets 128	*Potatoes in a Foil Pouch* 122
Mild Italian sausage 23, 194, 195	Pumpkin 180, 186, 187
Minced clams 35	*Pumpkin Cake* 180
Mini-Polish sausages 24, 26, 205	*Pumpkin Mincemeat Squares* 187
Minute Steak Skillet Supper 103	Pumpkin pie mix 180
Natural casing wieners 24, 69, 110	*Quartered Lettuce* 45
Old-Fashioned Meatloaf 82	*Quick Clam Chowder* 35
Orange Dessert Salad 44	*Quick Spanish Rice* 158
Orange-Pineapple Glaze 75	Red Delicious apples 49
Oriental Chicken Bundles 116	*Reuben Casserole* 94

Reuben Sandwiches 95	Steak and Shrimp Pinwheels 140
Rice 68, 84, 85, 89, 92, 116, 130, 136, 156, 157, 158, 159, 160, 163	Strawberry Pie 188
	Stuffed Cherry Tomatoes 14
Rice and Pineapple 159	Stuffed Mushrooms 21
Roast Duckling 71	Stuffed Peppers 84
Salisbury Steak 104	Stuffed Peppers - Italian Style 85
Salmon Patties 129	Stuffed Pork Chops 88
Sauces for Meatballs 197	Summer sausage 168
Sauerkraut With Bacon 170	Swedish Meatballs 20
Sausage and Cheese Casserole 98	Sweet and Sour Mini-Bites 26
Sausage and Chicken Gumbo 74	Sweet Potatoes and Oranges 171
Sausage and Pepper Bake 99	Sweet Sour Cabbage 93
Sausage and Potato Appetizers 24	Sweet-Sour Green Beans 167
Sausage and Sweet Potato Casserole ... 100	Swiss cheese 4, 94, 95, 113, 119
Sausage Cake 9	Swiss Steak and Dumplings 107
Sausage-Cheese Puffs 23	Taco Burgers 114
Schnitzel 108	Teriyaki Barbeque Ribs 123
Sea-Foam Salad 46	Tomato Sauce 197
Seven Layer Salad 47	Torsk 128
Shrimp (prawns) 14, 25, 126, 130, 140, 192	Tuna and Noodle Casserole 131
	Twice-Baked Potato Casserole 153
Shrimp Creole 130	Veal cutlets 108
Shrimp Spread 25	Veal Parmesan 108
Sirloin steak 57	Veal Shanks-Osso Buco 109
Sirloin tip 105, 135	Veal shanks 109
Skirt steak 140	Vegetable Stir-Fry 173
Slow-Cooked Lamb Shanks 78	Waldorf Salad 49
Smoked Cajun sausage 74	Whole Ham On the Grill 120
Smoked ham shank 32	Wild Rice Dressing 160
Smoked Polish sausage ... 97, 98, 121, 124	Wilted Lettuce Salad 50
Smoked pork butt 167	Yellow (summer) squash 173
Sour Cream-Dill Gravy 90, 193	Zucchini and Tomatoes 174
Spaghetti sauce 66, 73, 81, 108, 166, 194, 195	Zucchini squash 31, 62, 173, 174, 175
	Zucchini-Potato Patties 175
Spicy Wieners and Beans 110	
Spinach 13, 48	
Spinach Salad 48	

Made in the USA
San Bernardino, CA
17 February 2017